A Inman.

CW01653178

The author, Tom Buntin, on the unrestored Nick Stick Seat.

LIFE IN LANGDALE

LIFE IN LANGDALE

The memoirs of a Lakeland farmer

Tom Fletcher Buntin

Titus Wilson, Kendal
1993

© Tom Buntin 1993
ISBN 0 900811 24 2

Printed by Titus Wilson, Kendal

CONTENTS

ACKNOWLEDGEMENTS

I am most grateful for help received from the following people in Langdale and elsewhere – Anthony and Mary Divett for hours of study and for transcribing the old documents in my possession; Colin Whittle and Jim Birkett for their specialised knowledge of the plants, birds, and animals of Langdale; Janet Martin for sorting out the two Millbeck farms; the Armitt Library, Ambleside, Kendal Library, Cumbria Archives Service, Kendal, and the National Trust for making material available and assisting with research into the farm histories.

Professor R.N. Haszeldine and the Revd Douglas Stewart have read the text and made valuable suggestions, and many people have kindly allowed me into their homes or lent precious photographs, not all of which appear in the book: among them, Sid Cross, Vic Gregg, Molly Hodgson, Johnny Holmes, Jonathan Myers, Beattie Park, and Dot Parker. I am grateful to Kenneth Shepherd for permission to reproduce 'Storm Clouds over Langdale'. Last, but by no means least, my thanks to David and Mary Quine without whom this book would never have been written. David has transcribed my notes, typed the original text, made the maps and line drawings, and copied many of the photographs.

Tom Fletcher Buntin
Loughrigg Fold, 1992

EDITOR'S NOTE

It has given me very great pleasure to help Tom Buntin's book through the press, and I too would like to thank David Quine for his valuable help, and Oliver Turnbull of Titus Wilson & Son for much patient care. The original text has been slightly shortened and regretfully the chapters on flora and fauna had to be omitted in order to keep the price to a reasonable figure. I have made very few alterations, and the book is substantially as Tom wrote it. Langdale has so far lacked a historian, and Tom's work will hopefully inspire others to continue what he has begun here.

The publication of this book has been assisted by grants from the Curwen Archives Trust, and from another organisation which wishes to remain anonymous, with the co-operation of the Armitt Trust, Ambleside.

Janet Martin
Finsthwaite, 1993

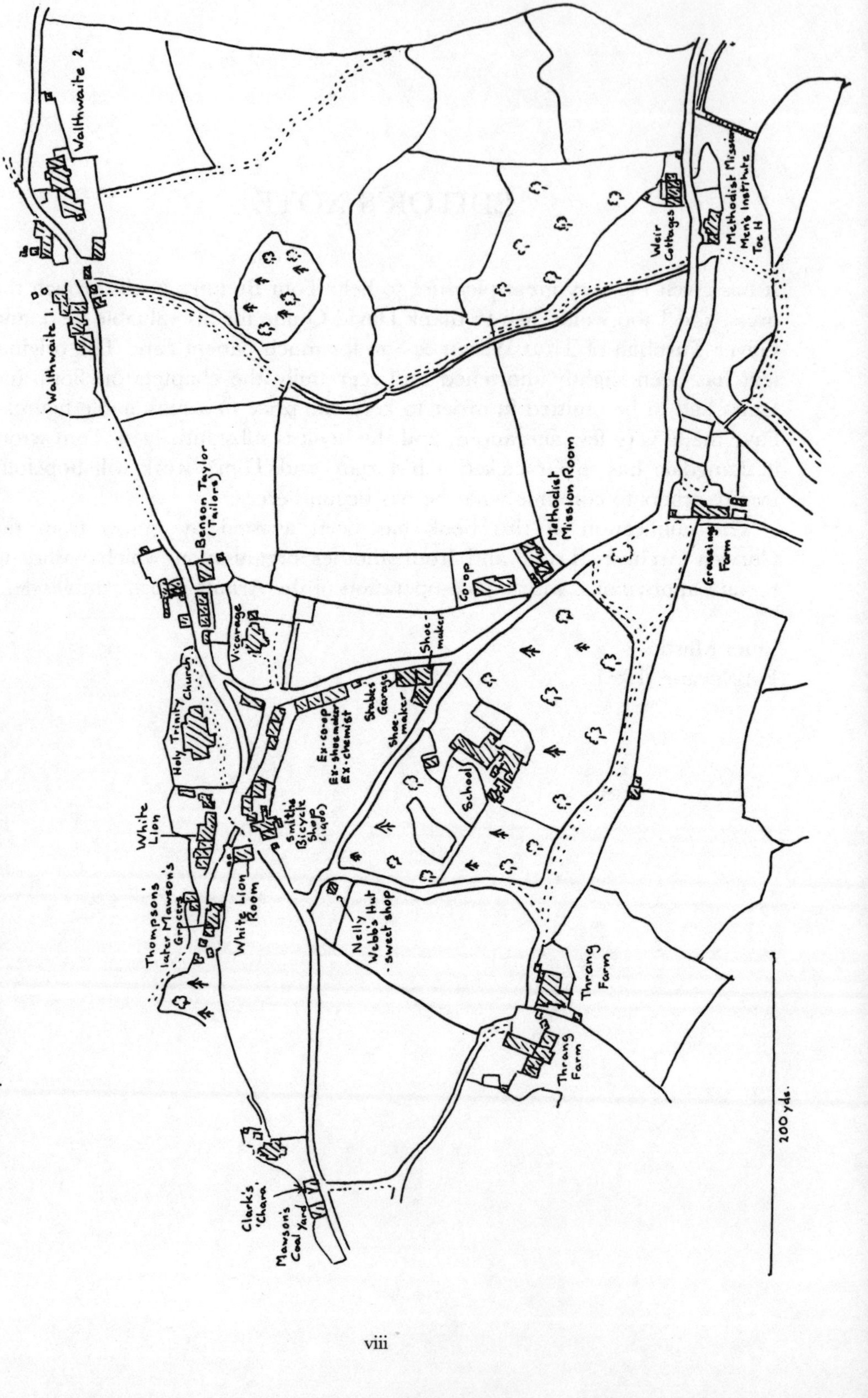

Chapel Stile in the 1920s

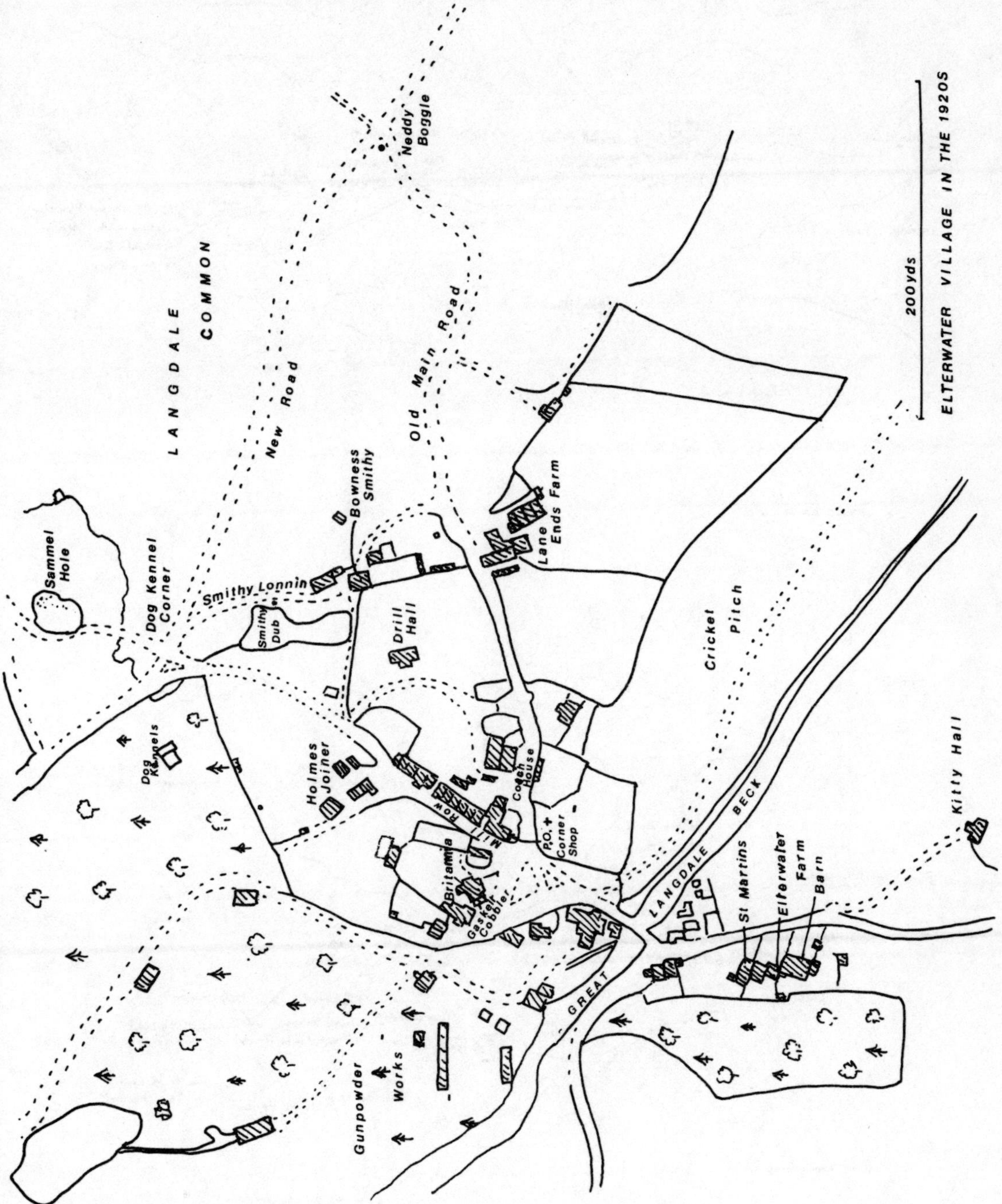

ELTERWATER VILLAGE IN THE 1920S

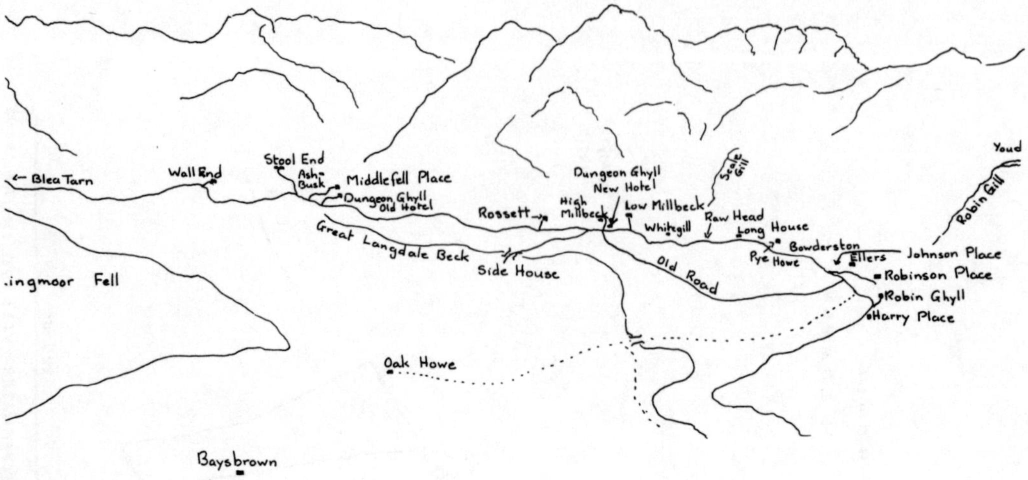

Great Langdale

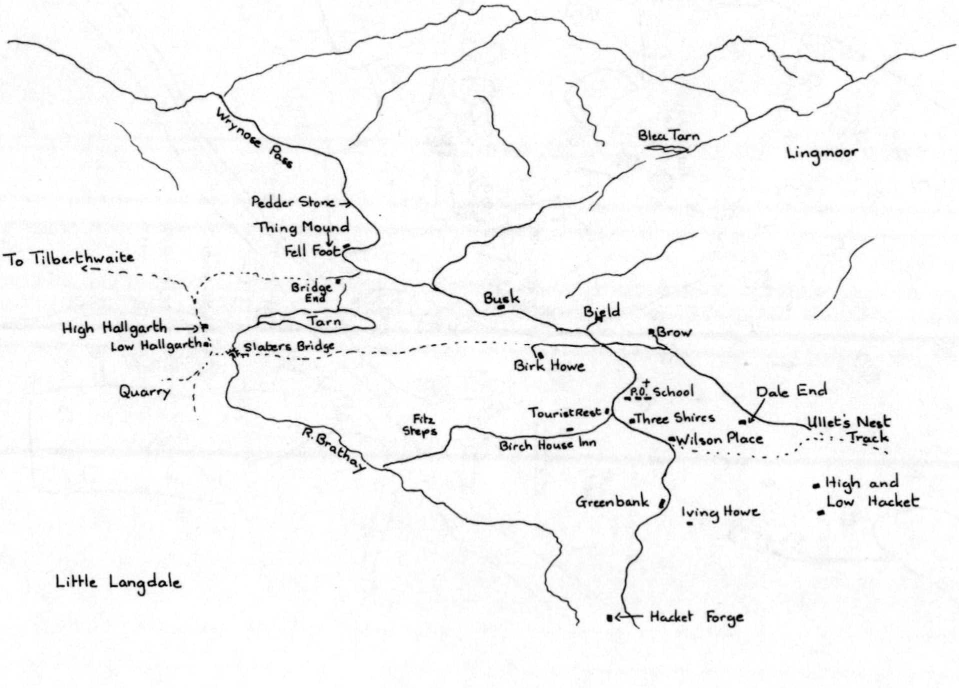

Little Langdale

CHAPTER 1

MY FAMILY AND LANGDALE

Family background

I was born at Pye Howe in Great Langdale on 15 February 1910 and have lived in the valley all my life. My grandfather, John Buntin, died the year I was born. He was a farmer and came from Grasmere but went to America in 1855 for a number of years, returned and married Jane Harrison. Both then went out to America, to Todds Point, Illinois, where they farmed with Merino sheep for 11 years. He then came home to Grasmere where my father, an only child, was born in 1874. In 1883 they bought Robinson Place in Great Langdale, where they lived and farmed. In 1896 they bought Pye Howe and the two farms were run as one. My grandfather was a good farmer, straight, honest, and responsible; he was churchwarden in Chapel Stile for some time and supervised the work on the new road from Skelwith Bridge to Elterwater. My grandmother liked farming and gardening and had a good orchard, often filling my milk can with plums or apples when I was a boy. She was also very tough, could quickly yoke up the horse and cart, and was prepared to spread muck or tackle other heavy jobs.

My father, John Fletcher Buntin, was a farmer, a conscientious person, quite scholarly and musical, playing the violin for pleasure and at local dances. He married Hannah Stables in 1896, they were born within two weeks of each other. My mother was delightfully unselfish and hard-working, milking the cows, mending, knitting, baking. I never saw her without some work in her hands. She left school when she was 12 years old and yet she was quite scholarly in her way and a very good pianist. My mother's parents, John and Jane Stables, were a well-known Langdale family living at Walthwaite in Chapel Stile; John had a little farm and a shoemaking business at Daw Bank. They had a family of seven children. Henry, the eldest, had a shoemaker's shop where the Post Office used to be in Chapel Stile; Thomas left for America when he was 16, working in a timber merchant's and married Sylvia Darrah; Agnes married late as she looked after her brother John; Hannah, my mother; John had a shoemaker's shop in Ambleside but died young. Jonathan married May Newby but had no family; he was an architect and a pioneer rock-climber, making several first ascents, without his father's knowledge. He was as lish (agile) as a cat. Mary, the youngest, died aged 19 years.

I was the youngest of our family. My brother John Fletcher was born in 1897 at Ellers; he was unmarried, quiet, reserved, a very good shepherd, and played the mouth-organ and the accordian for hours on end. Of my four sisters, Hannah was born in 1898 at Pye Howe, she was very strong-willed, unmarried, taught in the Sunday School for many years and was the church organist for 65 years, she used to walk to Ambleside for organ lessons; she would do anything around the farm. Margaret Jane was born in 1901 at Walthwaite, as my father was in America at the time. She sang in the church choir for many years, played the piano a little, was excellent with her fingers at needlework, crochet, knitting, and was a great traveller, visiting many countries including China when she was over 80. Esther Agnes (Hetty) was born in 1905 at Pye Howe, she was a very good pianist and had a beautiful singing voice and sang in the church choir. She married John Harris, the shoemaker who had the cobbler's shop in Ambleside and they had three children, Jeffrey, Gillian, and Sylvia Margaret. The fourth daughter, Sylvia May, was born in 1907, unmarried, and a very good violinist; she helped on the farm but was more like a boy, tinkering with engines!

I was the youngest. I played the piano and the organ for Hannah when she was off duty (but I like the cello best) and sang in the church choir. We played at home for pleasure, often in the evenings; Hetty, Sylvia, and I formed a trio and came second in the Wakefield Musical Festival Competition. We often played in local concerts and for dances. During the war, under duress, I was persuaded to play as pianist in the Santa Marina Dance Band, formed by some lads from Grasmere, their pianist was not available for some time. We played all over the place but it was difficult to work on the farm next day when I had only got to bed at 2 or 3 in the morning! I was not really cut out for farming but had to follow in my father's footsteps. I was particularly interested in everything to do with spinning, felting, crook-making, and local history, and in botany.

In 1954 I was one of the volunteers who recorded wild plants for the Botanical Society of the British Isles in the 10 kilometre squares which covered Great and Little Langdale, helped by Jim Birkett and Mrs Satow. The survey took me to all sorts of places, ghylls, woods, and even up rock faces. I remember going up Pavey Ark to see a plant called lady's fingers, *anthyllis vulneraria*. Jim had me on the end of a rope, it was a hair-raising experience for me as I have not a good head for heights. When the five years of the survey were ended, the records were compiled into a book, the *Atlas of British Flora*, published in 1962.

My father was an only son, a scholar, who was sent away to St Bees School and later in life, between times, he wrote articles for the *Lake District Herald* using the name of Robin Ghyll. He had numerous old papers and never threw anything away. He died in 1969 in his 96th year. A short time ago I moved house and from his desk and shelves I filled two tea-chests with 'rubbish'. Included were old newspaper cuttings, parish magazines, letters, my father's scrapbooks and diaries, which the mice had partly eaten. There were also old deeds and documents, some dating back as far as 1576 – 12 years before the Armada. These, together with my own diary complete from 1929, recording my own interests and observations, have enabled me to write this account of the Langdales.

Peaks and valleys

The two Langdales, Great and Little, are situated in the centre of the Lake District and are very popular with tourists. Now, with the motorways, a visitor can get up in a short time. The M6 comes to just south of Kendal where the bypass leads to a good road as far as Windermere. Leaving work early on the Friday night Langdale is reached in about two or three hours. One man reckoned that he had reached Little Langdale from north London in just over four hours. Travel back on Sunday night provides a nice weekend. Rock-climbing is a favourite sport of the young people and Langdale is a very good rock-climbing centre. The principal crags for climbing are Pavey Ark, Scout Crag, White Ghyll Crag, Gimmer Crag, Raven Crag, and Bowfell Buttress. You can be on the summit of Scafell Pike in quite a short time.

The Langdale Pikes, Harrison Stickle and Pike o' Stickle, provide some of the most spectacular scenery in the Lake Country, and are very much sketched and painted from just below Harry Place. When you get too near, you don't quite get the same impressive view of them. Another marvellous view of them is from Blea Tarn.

The Cumbria Way, from Ulverston to Carlisle, passes through Little and Great Langdale and is used by many walkers. Others prefer to be involved in the different fell races. In 1988 a new race was inaugurated called the Old Counties' Tops Race. It starts and finishes at Dungeon Ghyll and involves climbing to the summits of Helvellyn, Scafell Pike, and Coniston Old Man, a distance of 37-40 miles and a climb of 10,000 feet. In 1988 it was completed on one of the hottest days in June in 7 hours 41 minutes 8 seconds, the two brothers David and John Nuttall arriving together 20 minutes ahead of everyone else.

At the top or head of Great Langdale the valley is cut into two by the Band, thus making two valleys. The one to the left is Oxendale, starting from the farm of Stool End. It is a fine walk up there with a few nice ghylls. Kettle Ghyll, sometimes called Trap Ghyll, is on the left and comes down from Pike o' Blisco. Further up on the left hand is Brown Ghyll and then Crinkle Ghyll coming down from Crinkle Crags. Hell Ghyll which starts near the Three Tarns is a fine ghyll with spectacular falls of water. It is a lovely walk up the ghyll bottom if there is not too much water and a number of interesting plants and ferns are found here including the green spleenwort, *asplenum viride*, which is quite rare in central Lakeland as it is usually found on limestone. I have spent many enjoyable hours in there. Another very interesting plant grows by the side of Crinkle Ghyll – the interrupted club moss, *lycopodium annotinum.* It is a lovely walk up there in the autumn when the rowan berries are ripe.

The Crinkle Crags are a most impressive sight. It was on one of these, Gunson Knott, that an aeroplane crashed on 4 June 1937. This was an R.A.F. bomber. The two men on the plane, Pilot Sergeant Frank Wilkinson aged 25 and Leading Aircraftsman Alexander Mitchell aged 27, were both killed instantly. The plane was flying from Bristol to Donibristle in Fife where they were both stationed. It crashed in thick mist on the Friday, and although various people heard the thud of the crash, it was not found until late on Saturday. It was in a very dangerous place so the two bodies were not removed until Sunday night. Hundreds of people went up to see it. I went up on the Sunday morning with a Jessie Birkett who was only nine – she was staying with us on holiday at the time. It was quite unusual for us to have a fog like that in June – we had never seen the tops for three or four days. When the mist cleared and the sun shone, you could see the remains from the valley bottom.

Mickleden (large dale) is the right-hand valley formed by the Band and is about two miles long from Middlefell Farm. At the end of this valley is Stake Pass which rises to 1,576 feet above sea level and links with Langstrath and Borrowdale. In 1880 it was proposed to build a road over Stake Pass to connect the two valleys. Fortunately, the scheme was never developed.

To go over Rossett Ghyll, the steepest and roughest pass in the Lake District, you turn left at the bottom of the Stake Pass. From the top you continue to Esk Hause which is at 2,490 feet and links with Wasdale and Borrowdale. The direct, steep route up Rossett Ghyll has been fenced off to prevent further erosion. You can avoid the steep route by joining the old pony track to the left of the ghyll. At the top of Mickleden under Rossett Crag is the 'auld wivc's grave' – it is supposed to be the grave of an old pedlar woman.

In the visitors' book of the Old Dungeon Ghyll Hotel someone wrote the following poem:

> If I were a lover and loved a lass
> Who lived on the top of Rossett Pass,
> While I abode at Dungeon Ghyll;
> I swear by all that's good and ill

To love and cherish her ever and ever,
But to go up to visit her – never, oh, never!

Some 4,000 years ago Neolithic man in Langdale discovered a fine-grained volcanic rock from the Borrowdale series which he could use to make his tools. The band of rock comes to the surface at a high level on the Langdale Pikes from Mart Crag Moor to Harrison Stickle. In the rock wall of Pike o' Stickle is a man-made cave. The stones were selected and worked on site, leaving deposits of thousands of waste flakes and many roughed-out and rejected tools on the screes. The selected tools were then taken to the coast to be smoothened on the sandstone rocks.

Rain, tarns, waterfalls, and water-plants

The rainfall in the Lake Distict is a national joke and we do get more than our share – on the other hand, our rain is nice rain! The ground being light and stony, apart from the valley bottoms, the surface water drains away very quickly. The resulting green of the grass and the trees do, I think, make up for the weather. On a day when Barrow-in-Furness has half an inch of rain, Langdale has five times that amount. The head of Langdale, in Mickleden for instance, is one of the wettest places, apart from Borrowdale. However, further down the valley, even in Chapel Stile, the rainfall is very much lower. For instance we have been leading hay at Robinson Place when it was not possible at Stool End.

The annual rainfall figures show a marked drop as you go eastward, as much as 40% in less than 10 miles:

Mickleden	—	130 inches
Dungeon Ghyll Hotel	—	110 inches
Ambleside	—	75 inches

There are a few sheets of water in Langdale: Elterwater Tarn (elter = swan), Little Langdale Tarn, and Blea Tarn are the main ones. Stickle Tarn is situated just below Pavey Ark and was dammed and made slightly larger by the Elterwater Gunpowder Company over a hundred years ago. Loughrigg Tarn is another beautiful sheet of water where two water lilies grow, the white and the yellow, *nymphaea alba* and *nuphar lutea.* Up on Lingmoor, the hill dividing Great and Little Langdale, is Lingmoor Tarn which is small but a very interesting plant grows there – a bladderwort, *utricularia minor.* Another rare bladderwort, *utricularia ochroleuca* grows in some small sheets of water behind Raw Pike.

Youdell Tarn above Robinson Place has been made by cutting peat and the beck from this sheet of water runs down through the garden. We have skated on this tarn many times, but now it is full of reeds and bogbean so that the surface is not good for skating. It has recently been taken over by the black-headed gulls for breeding in the spring and summer.

There are a few beautiful waterfalls in the Langdale area. The most impressive is the Dungeon Ghyll waterfall which inspired Wordsworth to write his poem, 'The Idle Shepherd-boys, or Dungeon Ghyll Force':

It was a spot which you may see
If ever you to Langdale go;
Into a chasm a mighty block
Hath fallen, and made a bridge of rock;
The gulf is deep below;
And, in a basin black and small,
Receives a lofty waterfall.

This poem sums it all up, it is very narrow and has a drop of about 60 feet. It is impressive even when there is not a lot of water flowing.

Skelwith Force is very spectacular too, particularly when it is in flood. This fall is too high for the salmon to get up – no salmon are seen further up the valleys. It is quite a sport with young people canoeing down this river and shooting the rapids. Above the falls the water was used, by way of a sluice, to drive the mill at Skelwith.

Colwith Force is a waterfall worth seeing at all times of the year and, after a downpour, is one of the finest in the district. It consists of several falls accomplishing in all a descent of about 40 feet into the picturesque little glen. This used to be seen only by making a small payment at High Colwith Farm.

Great Langdale

Great Langdale has a long and comparatively flat bottom and there are now eight viable farms – years ago there used to be about 30 farms, some of which were very small and were amalgamated with others to make larger holdings. The valley bottom is quite good soil, some of red clay and some of a peaty nature and is capable of growing good crops of cereals and vegetables, but unfortunately the weather is not usually very good in the summer so that harvesting is a problem.

The old lane in Great Langdale, starting at Ellers, as it used to be called, was Raw Lane or Raw Lonning. The group of small farms went by the name of the Raw (= row), starting with Ellers, then Pye Howe, Long House, and the end of the Raw was Raw Head, as the name implies, the head of the row. The fellside above these farms is called Raw Breast with the highest point Raw Pike. High above Raw Head there are many little sheets of water showing where peat was cut. This was brought part-way down the fell to be stored in the peat houses which are still standing about 700 feet above the farms. Another little bog, where peat was cut near Youdell Tarn, was locally called Geordie Walker's Peat Pot.

Above Pye Howe is the sled road, which winds its way up the fell, we

The author's grandparents, John and Jane Buntin, in 1860.

The author's parents, John Fletcher and Hannah Buntin.

Shoemaking at Daw Bank, *c.* 1890; Harry Stables seated right, his father John behind, and grandfather Henry standing centre.

called it the Pye Howe sled gate (gate = track, road, way). This was the track used by Langdale folk when they went over to Grasmere with their cloth to the fulling mill. This was also the route taken by George and Sarah Green when they came over from Easedale to a sale held in one of the farms on Saturday, 19 March 1808. On their way back they went into a severe snowstorm and perished and were not found until the following Wednesday. They had left six children at their home at Blind Tarn, the eldest was only 11 years old. Nobody now seems to know the precise route they took when going home nor exactly where they were found dead.

Ellers was a rough-and-tumble small holding. In 1909 it was sold to the Revd William H. Sharpe, demolished, rebuilt, and renamed Ellergarth. There was a blacksmith's shop near the old lane which was also pulled down and the stones used in building the new house.

Below Robinson Place, where the old road heads up the valley, there still remains a stone seat, called the 'Nick Stick Seat'. By 1988 it was reduced to two flag seat stones, the right-hand one having a curved end is in its original position – the other stones had probably been removed to carry out repairs. The seat is backed by a 'single jammy' wall made with one thickness of cobbles from the beck, as opposed to a 'double jammy' wall with throughs built into it. A very ancient custom connected with the common rights of pasture was called locally the 'Nick Stick' meeting and was held annually on Holy Thursday. The lord of the manor's bailiff attended along with the tenants claiming turbary (the right to cut peat or turf on a peat bog or peat moss) and other rights on the common land. An agreement was reached for the people to pay the bailiff their proportionate shares. The last of these meetings for 'nicking on the stick' took place in 1888. The sum of £4.16s.4½d. was paid on this occasion to the bailiff. Towards the end of 1988 a team from the National Trust, led by Martin Scrowston, made a very good job of restoring the old seat.

Relating to this seat, my father wrote to Lord Lonsdale, on behalf of the commoners, protesting against the erection of a wooden hut by Mrs Ann Preston on the edge of Langdale Common, opposite the present Thrang Brow, without their consent. He received the following in reply:

Lowther,
Penrith.
26th. Febry. 1902

Gentlemen,

I beg to acknowledge the receipt of your letter dated 28th. December last protesting against the erection of a wooden hut on the edge of Langdale Common without having obtained the consent of what you term the "Common Owners".

I believe it is within the knowledge of some if not all of you that the Commoners who enjoy rights of pasturage on Langdale Common are entitled to such right on payment of an annual rent of £4.16.4½d. to the Lord of the Manor, which should be collected and paid over to the Bailiff of the Manor on the "nick stick" or stinting day, and that in recent years many of the Commoners, yourselves among the number, have omitted to make this payment and until they discharge this obligation, I am unable to see that they are in any better position than the Occupier of the Hut referred to. It was suggested at the last Court that a Meeting should be held to consider what could be done to improve the Common and settle this question as well as the question of encroachments and I should be glad if such a Meeting could be arranged when the Spring is more advanced.

I am, Gentlemen,
Your obedient servant,
Wm. Little.

To Messrs George Cragg, Henry Stables, James Stables, Joseph Youdell, William Ellwood Martindale, John Buntin, James Graves and J. Fletcher Buntin per Mr. J.F. Buntin, Langdale.

The National Trust is largely responsible for Langdale being preserved in its unspoilt state. Professor G.M. Trevelyan, Master of Trinity College, Cambridge, bought and gave to the Trust in 1929 the Dungeon Ghyll Old Hotel, and the farms at Stool End and Wall End, a total of 400 acres, in order to save the head of the valley from possible unsightly developments. Since then the Trust has obtained more of the valley and much of Little Langdale.

Across the valley from the Raw is the farm of Oak Howe. There is a strange entry in Hawkshead parish register in 1579 which is supposed to refer to a burial from Oak Howe: 'Anthony Dixson buried in Langdall the last day of December & taken up agayne & brought to Hauxhead the xjth day of January', was supposed to be a victim of the plague. Miss Armitt has this comment: 'a little plot of ground under bleak Lingmoor, ringed round with broken walls and overgrown with alien laurels is shown as the place where these people were buried'. They were most probably the victims of some pestilence. Relatives of these people at Oak Howe had sent some infected clothes from London and this is supposed to have been the source of the plague in Langdale.

The intakes (enclosures of common land on the fellside) in both valleys are mostly infested with bracken, *pteridium aquilinum.* The millions of spores given off by the bracken concern many because of the suspected connection with cancer in cattle and maybe humans. In this district it flourishes best in good soil and, growing to a height of 6 or 7 feet, it is becoming a menace as it prevents grass growing and is poisonous to cattle. They may die of bracken poisoning up to three or four months after eating it – there is no cure. When it is small it can be an advantage during a hot dry spell as it provides a little cover for the growing grass and prevents it from being burnt up by the sun. About 20 years ago a spray called 'Asulox' was perfected to kill it. We had our land sprayed by a light plane or helicopter in 1972. The kill was 85% effective, but if it is not cut or sprayed repeatedly the plants soon grow again. The spray has to be used towards the end of August just before the bracken starts to lose its colour. In the year after there had been a good kill thousands of foxgloves came up. These seeds must have been lying dormant for years. In years gone by the best method of bracken control was by cutting, and this had to be done in early July. As the old saying goes, 'Mow in June they'll come again soon, but mow in July, they'll very soon die'. Cutting with a scythe is a laborious job and in these days the farmer just has not got the time and hired hands are too costly. Fortunately, it does not grow much above 1,500 feet.

Some parts of the fells are covered with savin or juniper, *juniperus communis*. The clumps of savin are good shelter for the sheep in rough weather – they even eat it when there is a heavy snowfall. Unfortunately, the rough branches pull the wool off the sheep so that the clip of the wool then is not very good.

Camp sites, hotels, hostels, and climbing-huts

Camping is not allowed now on the National Trust farms. There is a National Trust camping site with a warden at the top of Great Langdale where they record 50,000 'camper nights' per year, and camping is allowed at Baysbrown. A very good caravan site has recently been established at Greenhowe, run by Mr John Holden and his son Graham. There was a Holiday Fellowship hostel at Wall End for many years, made up of about half-a-dozen wooden huts, but these were dismantled a few years ago.

At the top of Great Langdale is the Old Dungeon Ghyll Hotel, with the New Dungeon Ghyll Hotel about a mile down the valley. Drinks and meals can also be obtained at the Stickle Barn which has bunk-house accommodation. At Chapel Stile there was the old White Lion, next to the church; it used to be the house of the priest, who supplemented his income by selling ale because it was such a poor parish. This hotel licence was transferred in the 1930s by William Smith to what was the Grassings – the architect was Jonathan Stables, an old Langdalian resident. It was made into a very charming hotel but has been extended many times since then and became the Langdales Hotel. It has now been bought by the Langdale Timeshare Group and the pub has been renamed the Wainwrights.

At Elterwater there is the Britannia Inn, and a hotel, called the Eltermere, which used to be a private residence. In Little Langdale the Three Shires Inn used to be called the Tourist's Rest and was owned by Mrs Allonby. There was a beerhouse called the Birch House Inn on the right-hand side of the road to Fitz Steps but this was done away with many years ago. The Skelwith Bridge Hotel used to be called the Hare and Hounds Inn.

There are two Youth Hostels now, one made out of the farm buildings of the former Elterwater Farm; the second, at High Close, was a farm and then became the residence of Mr Jones-Balme. A Mr Wheatley-Balme, a wealthy businessman, came over from Mirfield in Yorkshire and bought High Close in 1857 from James Greenwood and made it into his private dwelling. He had a great interest in the area, he and Colonel John Robinson were the sole benefactors of Langdale church when it was rebuilt in 1857. He greatly enlarged High Close and developed a large garden. He built drives to Loughrigg Tarn Foot and the bottom of Red Bank all at his own expense. He was an off-comer but a great help to Langdale financially.

The farmhouse at Raw Head was bought and turned into a hostel for the Fell and Rock Club and some of the old buildings nearby became an annexe for visiting climbers. Another hostel, which is for Catholics, is called Bishop's Scale, and was made from the old farm buildings at Raw Head in 1944. The 'hoggus' (hogg house), a little further up the valley, was made into the Roman Catholic chapel at a later date. On the opposite side of the road a derelict cottage, possibly a small holding in years gone by, known as Raw Head Cottage, was bought and renovated by Miss Lamb of Caldy (for £500). This was in memory of her brother, Robertson Lamb of Bootle, who was very keen on the fells and died in 1927. She searched for two years before finding this old place in Great Langdale and named it the Robertson Lamb Hut. It is very small. The top half is fitted with bunk beds and the ground floor is for cooking and sitting. It is much used by the members and is a very good centre for rock-climbing. The National Trust own a climbing-hut adjacent to Blea Tarn House. There is another in Little Langdale at Fitz Steps.

Little Langdale

Little Langdale is the smaller of the two valleys and can be approached from the north by the Blea Tarn Pass, or better, from the south via Elterwater village and Colwith. There is an old road up the Ullets (= Owlets) Nest but this has gone into disrepair in recent years. The road up this valley is narrow and very wriggly, and as it is the through road over Wrynose and Hardnott passes, it gets quite chaotic with traffic at times. There are not many houses and a number of them are holiday homes. There are two small holdings at Colwith and four bigger farms up the valley, Birk Howe, Wilson Place, Dale End, and Fell Foot. The latter is at the foot of Wrynose Pass: 'stories of old pack-horse and smuggling days are told about this isolated spot, the last house on what was the main road over Wrynose toward the old port of Ravenglass' (Collingwood, 1932). Behind it is a flat-topped hump of ground, the remains of the 'thingmount' of Viking origin. This is believed to be where they held meetings of their parliament. The dalesmen of long ago met and discussed all serious matters pertaining to this valley and the area. Here all weapons were left at the foot of the mound and each and every free man, whatever his rank or status within the community, was free to speak his mind without fear or favour.

The Roman road from Galava, the Roman name for Ambleside, to Fell Foot and Wrynose Pass ran via Rothay Bridge to Skelwith Bridge, came up the hill to the south and then westward to the present road where the lane from the wood comes in. It passes down to Colwith, rather higher than the modern road, and south of Colwith Force to Stang End and Bridge End. Thence it went to Fell Foot and the modern road over Wrynose practically marks its course. Nearly half-way towards the summit of Wrynose, not much more than a 'girt lump of stean', is the Pedder (Pedlar) Stone where weary pedlars deposited their heavy loads on the flat-topped boulder. After a good rest they could face the long climb up the pass. It is on the left-hand verge of the present road about 100 yards above the fell wall. Just below the summit of Wrynose, to the right of the road, is the Three Shire Stone which marked the joining of the old boundaries of Lancashire, Westmorland, and Cumberland. Nearby is the once-flourishing iron ore mine where the ore was excavated and transported down to Hacket Forge in Little Langdale to be smelted.

Little Langdale is a good centre for the Coniston fells, Coniston Old Man, Carrs, and Wetherlam. Greenburn is the valley going off to the left to the site of an old reservoir and there are remains of the copper mines which have not been worked for years.

The valley bottom of Little Langdale is more undulating than that of Great Langdale but it has a lovely tarn. Just below the outlet is the beautiful old Slater's Bridge, one of the most photographed and sketched packhorse bridges in the Lake District. It is extremely narrow with a low parapet, the slightly pointed arch being a characteristic of the oldest bridges in Lakeland. Here is a place to linger.

If you cross the bridge towards the quarries and turn left along the old road and then take the gated track up on the right you come to the tunnel at the quarry entrance. This long dark passage is filled with water. Wellingtons are a must here. The tunnel opens out into a fantastic cathedral-like dome with one beautiful shaft of rock supporting the roof. At its foot is a clear, deep, still, green pool. It is silent, except for the incessant drips of the falling water from the roof high above. Nearby, a different exit leads to a vertical rock-face which provides ample scope for those who enjoy trying out their abseiling techniques.

A local quarryman, Jim Birkett, the son of Tom, is a well-known naturalist and rock-climber. All the farmers who had a sheep crag-fast always went to Jim for help. I have seen him in action on Tarn Crag above Easedale and on Yew Crags. He would skilfully work his way up to the sheep, catch it, put a rope round it and lower it to the farmer at the bottom of the rock. He has no fear and will never leave a sheep in difficulties, commenting 'I wouldn't have liked to have been left in difficulties myself'. He is extremely knowledgeable about the flora and fauna of the area. Few rarities escaped his notice including visiting snowy owls and ospreys; he also studied the breeding peregrines and is aware of the haunts of the golden eagle, but is wisely reticent in divulging information which might be wrongly used.

Mountain rescue

In the 1920s there were only small numbers of visitors, and if someone was injured or lost the local shepherds and farmers would combine together to search for them and bring them down. I well remember seeing them and hearing about their rescue operations – there was 'Zeke' Myers of Wall End, Herbert Grisedale of Middlefell, Tom Birkett of Side House, Art Askew of High Millbeck and Thomas Youdell of Low Millbeck, Bill Parker of Long House, and Fletcher Buntin of Pye Howe. They were always ready to help out when need arose. Later Sid Cross and others organised rescue teams from the Old Dungeon Ghyll Hotel for many years.

Since so many people come to enjoy the Lake District nowadays more accidents do happen. The Ambleside and Langdale Mountain Rescue Team was formed by combining the Ambleside and Langdale teams in 1970. It is one of the busiest teams in the country with 80 rescues a year. It is purely voluntary, and the members who give so much of their time and energy have saved many lives. Perhaps surprisingly, out of 60 recent accidents only three involved rock-climbers. The team is alerted by its individual bleepers, and the Landrover aims to leave Ambleside within four minutes. They now have a new ambulance and can call upon the Sea King helicopter to come from Bulmer in time of emergency. Dogs are also involved in searching for people who are lost or missing. The Mountain Rescue Team has 40 persons; usually about 20 come out for any rescue. They do need support as in 1988 it required over £40,000 to maintain their services.

On a more cheerful note, the late Richard Clough, who retired to Langdale, had many gifts and wrote several poems which he had published including this ditty, titled 'Overweight':

If you feel you're too fat to climb Gimmer
It's time that you tried to be slimmer;
But to halve what you eat
Or put glue on your feet
Would only make Gimmer look grimmer.

CHAPTER 2
STONES, WALLS, AND BUILDINGS

Stone walls, hogg holes, and stoops

There are miles of stone walls in Langdale. The National Trust is responsible for 2,000 miles of walls in the Lake District, some dating back as far as the 13th century. In the early days the men who built walls did not mind working 10 or 12 hours a day; they must have been very strong as some of the stones they moved must have been two or three tons in weight. The stones at the southern end of Great Langdale are much easier to use for walls since there is a seam running across the valley which provides stone which is suitable for riving (splitting) into slates. At the top of the valley the stones are rounded – what we call 'beck cobbles', and when you were walling with these you had to put two rows of 'throughs' to bind the two sides together. A 'through' is the stone which goes right through the wall; it does not matter if it sticks out at each side. It is interesting to find stones differing so much in such a short distance. Just behind Raw Pike there is an outcrop of flint, which indicates a sign of copper, and at the top of White Ghyll a little iron ore.

The stones on the top of the walls we call 'cams'. My father taught me to wall and he used to say 'Always put the stones end in and end out, not trace ways, if you want them to stand'. 'Trace ways' is parallel with the wall. One side of the wall should be about 'plumb' and the other side with a bit of a 'batter' or inward slope. Therefore, if you start the wall at the ground 2 feet wide, it should be about 1 foot 9 inches at the top. It is most important to keep the middle filled with little stones called 'fillings'.

There are lots of 'hogg holes' left in the walls. These were holes for letting the sheep through from one field to another, and could be closed by a slate slab kept nearby. They are not so much in use in these days. Very often above the hogg hole would be a stile or 'stee' for the farmer to get over the wall since it was a tight fit for a big man to get through a hogg hole – now the stiles are mainly made of wood. The gateways were often made from a slab of stone, a 'stoop', with holes in it through which lengths of wood were inserted. These wooden rails were obtained by cropping ash trees, but most of these have been replaced by gates. Ash wood was also used for fencing-posts, but now many of the ash trees have died. Most farms had sufficient trees for their own use and two or three of these would be cropped every year and the rest about every seven years – they were done in rotation. If the trees were in a fence or 'dyke' they would be laid about every seven years.

Many of the fields were open to the road and there were gates across the roads but most of these have been replaced by cattle grids. One gate remained up by Blea Tarn until 1988. Among others were those at the top of Langdale Fell where Elterwater Common begins, and at Harry Place which we, as children, would open for the horse 'charas' and the people threw out pennies for us.

Farmhouses

The old Langdale farmhouses were mostly built on the same pattern – squat, square-shaped buildings, most with flattish square chimneys. There are one or two exceptions, and both Harry Place and Baysbrown have the tall rounded chimneys. Most of the farms had the barn or shippon joining at the end of the building and some even had a door out of the shippon into the house; this was originally so at Robinson Place. There was a back door into the kitchen and the front door into the living room and often a parlour on the right or the left. The dairy was always on the sunless side where it was cool. The walls of the farmhouses were often 3 feet thick and some of the stones were very large. The folk who built them must have been extremely strong. They would no doubt use thick pieces of wood to roll them into position – sometimes taking a great part of the day to place two or three stones. Time was not taken into consideration as they had a great pride in their work.

When a house or barn was built the roof timbers were hewn out of trees with an axe and the roof was covered with slates, some as much as 2 inches thick, called 'battens'. They were fixed to the timbers with wooden pegs. The roofs were very crude; they were not pointed on the inside so that you could see holes between the slates. If it was a wild snowy time the snow would blow through the gaps so that people would wake up with the bed covered with snow. At the top of the roof were 'wrestler slates' with bits chipped out so that they hooked together. When I was a lad there were some on the old barn at Pye Howe. Later ridging tiles were hewn out of a sort of limestone or some other soft stone. Now they are all fixed ridging tiles. The old guttering was made out of wood and I can recollect my father taking the wooden ones off Pye Howe and replacing them with the cast iron ones. Window lintels and frames were made of riven wood and pegged. An excellent example is in the middle or bracken barn at Robinson Place.

The chimneys were all very well built with no lead flashings, as that would be very expensive even if you could procure the lead. They were always done with stone flashings and kept nice and dry. All the chimneys were open so that you could stand at the bottom and look up at the stars. They had crooks a little way up where bacon and mutton was hung to be smoked. Coal was not used; it was peat smoke that was suitable. The peat fires were never let out as matches were scarce. I remember my mother talking of someone whose fire had gone out so they had to travel about a mile with a candle lantern to get a light but when they got back home the candle blew out and they had to go all the way back. We used to go out to the byre to milk with a candle lantern, quite often it was blown out, so we would hold it under our jackets for protection. In my younger days we had no other means of light, no torches or electricity.

At first the floors would simply be earth. Later slate flags were used of the Brathay blue type. Many were quite highly polished, but some were quite rugged. Many of the old flag-stones are still in the houses. Most of the old farmhouses had a spice cupboard let into the thickness of the wall near the fireplace to keep dry the salt and household spices. Court cupboards or bread cupboards, large enough to keep a month's supply of bread, were built in many of the houses, forming part of the partition between one room and another. They nearly always have initials and a date – the one in my present home is dated IB 1691, for John Benson. They are usually made of oak, carved and highly polished. At Walthwaite there was a real beauty, made of yew, giving it a reddish colour. Unfortunately, like many others, it has been taken out and sold. There were also a lot of oak partitions for the divisions of the bedrooms and the doors were of oak. Many of the partitions have been taken out or covered over with plaster. In the olden days the walls would be dry, but later they were covered with plaster which was called mortar and was made of lime and sand, no cement. The walls were given two or three coats with the last one being of very thin lime.

Baking was done on the hearth with a griddle and also on a bakestone or 'baksten'. This was a metal sheet built on a frame which supported the fire which was often fuelled with dried bracken. It was used mostly for making 'haver bread' and oatcake. Haver bread was made with oatmeal mixed with water and rolled out very thin (haver = oats). Oatcakes were not rolled out as thinly.

Beds were four-posters with valances hanging from the top and round the bottom. Wooden laths formed the base, on top of which were two straw mattresses and then a feather bed. This was turned and puffed up each day giving a very comfortable bed. Sheets and blankets went on top and a fancy quilt over all. Some of these quilts were beautifully made by the women of the house on a quilting frame, often designing marvellous patterns. The women also made the pegged mats put by the bedside to cover the bare boards in the bedrooms. Old cloth garments were cut into pieces and pegged into a piece of hessian. The peg was made from a piece of very hard wood, such as yew or savin, and polished very highly to make it easy to push through the hessian. These pegged mats are coming back into vogue. They are very nice and warm to stand on, the only disadvantage is that they hold a lot of dust.

The dairy, on the cool side of the house, had stone flags all around forming a working-surface on which the milk was stood to cream. There had to be sufficient room for three or four bowls – if the milk was put into bowls or tins on the Monday morning it was creamed on the Tuesday night. The cream was skimmed off with a skimmer and kept for a week or so before being churned by hand into butter.

Byres and barns

The shippon or byre was paved with rounded stones of about 3-4 inches. It was quite an art to lay them down. You wedged them together and ran watery gravel between them. When laid properly they were very hard to move. Now most of them have been replaced by concrete. In the byres there would be calf 'hulls' or pens, and the cows were tied up in stalls called 'boo-us' (booses). The old shippon at the end of the Old Dungeon Ghyll Hotel, now the bar, still has the stalls where the cows stood. The cows were tied up with a band passing through an iron ring and swivel called a 'red widdy' to a stake, called a 'red stake'. Now cattle are fastened with a metal red stake or run loose within the shippon or the calf hulls. This method makes it necessary to dehorn the animals to prevent them damaging themselves and others. In the shippons there was the 'fodder gang hole' – a trap-door – through which the hay was pushed from the floor of the barn above into the 'fodder gang' below to be fed to the cows. The 'fodder gang' was a gangway or passage made from slate slabs in a vertical position, held in place by a groove in a horizontal wooden rail. The hay was dropped down onto the floor ready to be lifted over the partition onto the floor of the 'boo-us' to be fed to six cows on each side of the gang.

There is not much milking done now in Langdale, and the farmers concentrate on sucklers. We always milked by hand, sitting on a wooden 'milking copy' or stool. It was quite a laborious job; not many young folk can milk by hand today. The womenfolk also learnt to milk which was very useful when the men were away. We learnt to milk when we were 12, often having to help before going to school. Visitors often wanted to learn to milk, but they didn't have much idea. We had to muck out with a wheelbarrow and shovel.

Some of the barns had the main beams running right to the ground, called 'crucks'. Good examples are at Rossett and Wall End, at the foot of the Blea Tarn road. The barns were often set into the hillside, the shippon being at the bottom with the barn over the top so that the horse and cart could be backed into it. These are sometimes called 'bank barns' and in Britain are only found in the Lake District and Yorkshire.

The carts were made of wood with wooden wheels encased by an iron hoop. If the ground was steep a sledge or 'sled' was used. We used them to get the bracken down for bedding from the intakes in autumn. The intake, land taken in from the common, was on the steep fellside. There are many little intakes in Langdale and the main purpose of these small walled enclosures was for cultivation. The ground would be cleared of rock and stones, some would be ploughed and used for growing oats, barley or 'bigg', and some vegetables.

The farmers in the very early days would have to be self-supporting, requiring a lot of effort, especially as the tools were all very primitive. In my young days the hay was cut by the horse mowing-machine, then we would cut round the edges, which were out of reach of the machine, with a scythe or 'ley'. There was quite an art in sharpening one of these without cutting your first finger. We used a 'strickle', a sharpening-stone attached to the end of the wooden shaft of the ley, so that you carried it along with you. Some of the very early leys had a blade about 6 feet long so that if the farmer was a short man he had to stand on his toes to reach the end of the blade to sharpen it! In the old days, before my time and before the horse mowing-machine, a swathe about 6 feet wide was cut with one of these leys so that six men in a row could get on quite quickly to cut a whole field.

Another very necessary tool was a peat spade. Many days were spent cutting peats and stacking them to dry. Later they were stored in the peat houses, such as those above Raw Head.

Around the farm

Outside the house there was always a 'sconcon' or 'seatson' for putting the buckets under, also to sit on. There was another stone called the 'horsing stone' which helped you to mount the horse. There was one at Walthwaite until recently. Houseleeks, *sempervivum tectorum*, were grown over the door as they were supposed to keep the devil away – Bridge End has a very healthy crop. It has very fleshy leaves which when they were rubbed on warts were believed to remove them. Another name for it is 'Welcome-home-husband-however-drunk-you-be'.

Nearby you would always find a 'bee house' or 'bee bole'. The bees were all kept in straw skeps so they needed to be under cover. The bee boles are mostly walled up nowadays, but you can still see where they have been. We had one at Pye Howe, and the shelf-like structure on the end wall of Raw Head was another. Wooden hives now do not need the cover of the bee houses. Very few bees are now kept but in olden days honey was used for sweetening.

A well was always to be seen, giving lovely cool water. The old earth-closets were usually built some distance from the house and often could accommodate two or three people. They are still in evidence but are now mostly made into tool sheds. Ashes were put down frequently; the closets were cleaned out about twice a year and the contents spread on the land – not a pleasant job, but it had to be done!

At the back of the house was the 'parrock', a walled enclosure where weak lambs or ewes could be taken care of, close at hand so that you could keep an eye on them. You could keep your horse there if you were needing to use it first thing in the morning. At Harry Place was the Hen Parrock, a small triangular field to the east, just below Robin Ghyll and bordered by the present road.

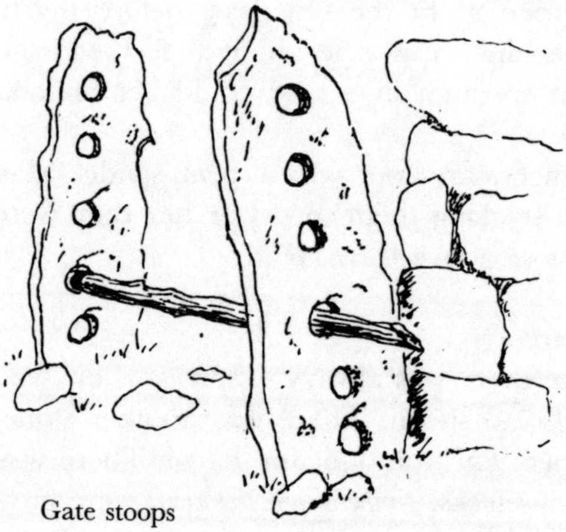

Gate stoops

CHAPTER 3

ROADS – OLD AND NEW

The Romans built their roads to a width of 24 feet in some of the valley sections, and to a remarkably high standard of workmanship. I have already mentioned the routes out of the valleys following the Roman road out of Little Langdale over Wrynose. Another old track goes off to the left just beyond the Three Shires Inn and leads down to the ford by the footbridge, then through the woods to Tilberthwaite. Fords were quite common; there was one just below Robinson Place which forded Langdale Beck and gave access for the farmer from Harry Place to reach his field called Rutts. To proceed out of Great Langdale you could leave by the pass by Blea Tarn or the packhorse route up Rossett Ghyll to Wasdale or by the Stake Pass to Borrowdale.

Originally the roads were unfenced and wandering sheep or cattle could damage growing crops. To counteract this walls were built to enclose the road. For instance, the road beyond Slops Bridge (Rossett Bridge) remained unfenced until the early 1920s. Just below Robinson Place, where the old road crosses the present bridge, the fields on either side were unfenced until my grandfather carted stones from the quarry to build a low wall with a wire on top at about the turn of the century. The field on the right was known as 'Carsa' (Causeway) and the wall was built to keep sheep and cattle from straying onto his fields.

Making the old roads – the stone-breakers

The roads in the two valleys are narrow and winding and were just made of stone and gravel until the early 1920s when tar-spraying came into fashion. When they were dry, clouds of dust would rise from them. The method of making them in the early days was to use stone chippings and 'sammel' or gravel. Big stones, either quarried in the neighbourhood or collected off the land after being turned out during ploughing, were carted by the local farmers to the various wide places of waste ground and then the stone-breaker would come and get to work. I can still see the old stone-breakers. Walking along the road in almost any part of the Lake District you could come across a large heap of stones. You would hear the ring of the stone-breaker's hammer long before you reached the place. When you drew near you found an old man seated on a sack, his legs stretched out before him and his trousers hitched up and tied with a piece of string below the knee – 'Yorks' we called them. He wore a pair of goggles to prevent chips of stone or flint from flying into his eyes. Often we would stop to bid him the time of day and he would lift his head for a moment for an exchange of greetings. He would linger for a break then go on with his work, talking at the same time. The stone-breaker was quite an interesting man to watch and talk to. He would tell you that although the work was hard, he would rather be out in the open 'brekking staans' than doing nowt at all! When he had broken up a heap of stones they were measured and he would be paid by the cubic yard.

The stones were broken to about the size of a tennis ball. A layer of these to a depth of about 5 or 6 inches was put on the surface of the road and then covered with gravel and watered in. The watering-cart held about 200 gallons and it had shafts to be yoked to the horse. Then a heavy steam-roller was used. The driver used to bring his sleeping cabin drawn behind the roller. He had to get up early to get his fire going to get up steam. The gravel was dug out of various sammel holes in the valleys. There was a good one on the present road below Robinson Place where the old valley road joins it; on the first bend of the old road the stone-breaker would be at work. The best sammel was the very hard stuff; it took some digging and the pick had to be sharpened frequently. As children we used to enjoy watching the steam-roller going backwards and forwards – and talking to the driver.

The roadsides would be cleaned out periodically by the farmer with the fields adjoining that part of the road. We used to cart the rubbish into a corner of a field and along with the debris we got from the pools and gutters, it made a good loamy soil. We mixed this with cobbled lime – when it was slaked we mixed and spread it on the land. This we called lime and mould. It was very good for improving the land.

Making the new roads

1. Langdale to Borrowdale via Stake Pass
   The *Ambleside Herald* for November 1880 reported:

   > A long discussion took place as to the practicability of a proposed road over Stake Pass. Ultimately it was resolved that the committee considers it desirable that a conveyance road be made over Stake Pass from Langdale to Borrowdale and that Messrs Brown, Garnett and Bowness be appointed as a sub-committee to consult with the likely parties as to the possibility of raising the necessary funds.

I cannot find any other papers on this – it was obviously never carried out!

2. Longbrow – Elterwater to High Close – *c.*1861-98
   Part of the road between Elterwater Common and High Close, known as

Longbrow, was steep and rough making it difficult for horses. A stretch of about 300 yards was improved by re-routing it to the north on a less steep gradient. It is still possible to see where the old road wandered its way up the common. This was done privately by one of the Balme family who owned High Close. They also developed the road down Red Bank towards Grasmere.

3. Langdale to Skelwith – 1896

The new roads were made over a period of many years. The first major improvement was made by the Gunpowder Company in 1896. It started in Elterwater at the old Police Station near Mere Syke and joined the Coniston road at Skelwith Bridge. The purpose was to make it easier for the horses which brought material from Windermere Railway Station to Elterwater. The road came round by Loughrigg which is still very steep in places. Before reaching Skelwith Bridge there was a building at Halfway House (reckoned to be about halfway from Windermere to Elterwater) where they could unload some of their heaviest burdens.

The county surveyor, Joseph Bintley, was consulted about exactly how the road was to be constructed, since it would be public, not just for the use of the Gunpowder Company and its horses. Three inches of stones, 'broken to go through a two inch ring', and covered with a 'good coat' of rolled and watered gravel were specified, together with sufficient drainage. My grandfather, John Buntin, was the supervisor of the work, and Joseph Bintley came several times to inspect it. On 9 March 1896 he reported that, although generally he considered that the work had been done 'in a substantial and workmanlike manner', some of the recommended drains had not been installed, there was an imperfect culvert, and some of the metalling had not been laid. Unless these defects were put right, he added, he would be unable to grant the certificate necessary for a public road, although he was of the opinion that the new road would be 'a decided advantage to the neighbourhood over the existing main road (Ellers Brow to Mere Syke) ... the gradients are easier and it is wider and generally will be more useful'. The new road was 1 mile 202½ yards in length, 12 yards in Langdale township, and the rest in the township of Loughrigg.

4. Road through the school ground at Chapel Stile – 1895-98

A new link road was proposed to go through the school grounds, then known as Brig (Bridge) Howe school, in Chapel Stile (the 'new bridge' between Thrang and Baysbrown had been built in 1818). This would avoid the hill in the village and the sharp corner by the church and run from the former Post Office to the Thrang road end. Again my grandfather was involved, the cost of £70 being raised by private subscription. I have the specification in my possession:

> The soil and trees to be removed and laid in the adjoining land. The wall on the west side to be two feet six inches wide at the bottom, with rows of throughs two feet in height and not more than six inches apart with discharge holes for the surplus water, and to be built five feet high above the finished surface of the road with a cam not less than ten inches deep. The road not to be less than eighteen feet wide. The hill on the north end to be taken down at least four foot. The rock by the school gate to be blasted out to form an even gradient. The surveyor will appoint out where the gates are to be put. The whole of the work to be done under the supervision and the satisfaction of the surveyor and to be completed before the 20th. March 1895. The road to be covered with sammel to the depth of four inches.

This road was completed in time to appear on the Ordnance Survey map of 1898.

5. Neddy Boggle to Dog Kennel Corner – date not known

Another new piece of road was made at about the same time on Elterwater Common starting at 'Neddy Boggle', where it turned down into Elterwater village, cutting straight through to Dog Kennel Corner. This stretch of road joined the new piece which the Gunpowder Company made up to Weir Cottages. The original Ambleside-Chapel Stile road went through Elterwater village, behind the Britannia Inn, alongside the beck on the piece of common which the Gunpowder Company took in when they enlarged the factory, and so into Chapel Stile. There is a guide stone at Neddy Boggle which was reputedly so named (Neddy = donkey, boggle = ghost) because every time a donkey was taken that way, when it reached the corner it 'took steck' and refused to go forward – it would have to be cajoled into moving with the help of a carrot! Dog Kennel Corner acquired this name from the kennels which housed Colonel Robinson's pack of hounds. Beef for the hounds was hung on trees on a wooded knoll, now in the timeshare grounds, called Beef Hill.

6. Road to Dungeon Ghyll – completed 1925

This new road, built at a cost of £7,500, started just below Robinson Place and followed the old Raw Lane as far as Raw Head. The final section from just behind Raw Head was cut through the fields to join the old road just below the New Dungeon Ghyll Hotel. Following the old lane accounts for the winding nature of the road. This road was made because the original one up the middle of the valley was so often flooded, sometimes to a depth of 2 to 3 feet.

The authorities brought along a crusher, which was driven by a traction engine to break the stones for the road metal – it created a great deal of noise. They employed a lot of men which was fortunate at the time as the Gunpowder Works were in a poor state and local men were on short time.

Ellers, Great Langdale, before 1909.

Low Millbeck: Herdwicks behind the farmhouse.

A four-horse chara at the old Dungeon Ghyll in 1910.

Huntsmen, hounds, and fox in the yard at High Millbeck.

They did various jobs, including walling, which accounts for some of the walls not being very well built, since many of the men had never done any walling before. It takes years of experience to become a good dry-stone waller.

The council had considerable difficulty in draining the old road and many problems in making the new one, as reported in the *Westmorland Gazette* on 16 May 1925:

> As is well known, rock abounds in the district, and for the excavating work experienced men have been employed. The large boulders which have to be removed are so hard that rock drills will only 'stand up' for a couple of inches. These boulders have been placed in full line in the old fences at some time, but today they cannot be removed without blasting. The track of the old road is only 7ft wide and the proposed width of the new road will be 20ft.

Not many years after it was completed a steam-roller tumbled down into the field and rolled over onto its side, just opposite the gate to Long House. The retaining wall collapsed, not being sufficiently well built to take such a heavy vehicle. However, it was a great improvement on the old road, particularly now when we have so many cars and tourists. The old road up the valley is very useful for walkers and gives the farmers access to their fields.

At the same time three new bridges were built, one at Millbeck to go over Mill Ghyll, the next at Rossett Bridge (or Slops Bridge as we called it) to do away with the hump-back bridge, and the other at the Old Dungeon Ghyll Hotel near Middlefell Place. These were all big improvements.

The Lengthmen

Around the 1930s there was always a 'lengthman' – he had a block of road which he was responsible for keeping clean. Tommy Birkett did Little Langdale, Ben Kirby, who had a hunchback, did Skelwith. Chapel Stile and up the valley was done by Will Birkett of Stonebank – I remember him stone-breaking in his spare time – then by Thomas Youdell, and more recently by Charlie Laight. Thomas Youdell knew every grill, little hole in the wall by the roadside, and every drain up the valley. On wet days, with his oilskin trousers and overcoat he would cycle the full length of the valley and clean out each one – every hole would be cleared of leaves. You never had water standing on the roads in those days! The job was no longer done after the end of the Second World War.

CHAPTER 4

SHEEP AND SHEPHERDING

Farming is a very interesting job and I would not have exchanged my life for any other calling. It is hard work and sometimes you have big losses – I think it has to be in your blood to enable you to keep going.

The year passes by with the rotation of tasks: spreading the manure on the fields, brush- or chain-harrowing, gutters to be cleaned out and blocked drains to be opened up. Brambles need to be cut back as they grow too long and the sheep get caught up in them; hedging must be done before the sap starts to run. Walls constantly fall down and must be put up as soon as possible or the sheep get through the gaps and make them bigger. These jobs were always done in the winter and the early spring. Then there were the bigger tasks at lambing time, and at clipping and dipping, plus hay time if you had any – some farmers buy it now. Not much ploughing goes on now, but it is useful if you have the time. There is never a dull moment on a farm.

The early days – intakes, tenancy

In years gone by farming would be the only industry in Langdale. Some of the farms would be very small, just a dwelling and a few fields. Gradually more ground was cleared and fenced in with walls. In a lot of cases the walls would be made with the stones collected when tilling a new piece of ground. This hillside land was taken in and the new enclosures were called the 'intacks' or 'intakes'. The bottom land would have a lot of scrubby trees and boggy areas, so it had to be cleared and drained. On some of it open cuts were dug and lots of old-fashioned stone drains were made – these were very efficient and many remain today. Some of these drains, with slate roofs covered with turf, are out of sight and go the whole length of the smaller fields.

The breed of sheep kept was the hardy Herdwick. There are two theories as to where this breed came from: one was that they came off a wrecked ship from Spain; the other that they came with the Norse settlers. I think the latter is the more likely origin. These settlers would also keep a few cattle and possibly some pigs and a horse or two. So with the sheep for meat and wool, and a crop of oats and possibly some vegetables, together with lots of edible wild plants, they would be mostly self-supporting.

Later on when these folk became more established they rebuilt their own dwellings, a hard task. Some of the old houses were built as far back as the 16th century. The womenfolk did the spinning and weaving and then they carried their cloth over the tops to Grasmere to the fulling mill there. Later a fulling mill was built at Millbeck – a corn mill stood there too – there was also one at Skelwith Bridge. As these mills needed a good head of water these were two good places for them to be.

It was (and is) customary for each farm to have a stock of sheep attached and let with the farm. Some of these stocks were quite small, in some cases not more than 50 to 100, while some of the bigger farms had 700 or more. When a farm is let to a new tenant, there is still the old custom of a 'livering' (delivering). This is very necessary as the tenant leaving the farm must leave the same number of sheep of the same ages and groups as when he took them over. Therefore, both tenant and landlord must have their 'viewers'. An example of an old viewing paper reads as follows:

A schedule of sheep delivered at Harrison Place on April 26th 1799 to John Greenup

Hoggs	39	– 26 very decent, 13 small (two of these disordered)
Twinters	35	– very decent
Wethers	29	– very good
Ewes	47	– with lambs or had lambs, a part of them bad.
Total	150	

In years gone by, an ingoing tenant had to find a substantial bondsman to guarantee the safe redelivery of the sheep. If any of the different classes of sheep were depleted it fell to the bondsman to make good the loss. Frequently the demands on the bondsman were very heavy. On 15 April 1735 the following agreement was drawn up:

> Know All men by these presents that wee Robert Jenkinson of Heaning house in Seathwayte in the Parish of Kirkby Ireleth . . . and Edward Benson of Millbeck in great Langdale . . . are held and firmly bound unto John Dixon of Rosffitt [Rossett] in Great Langdale . . . Concerning the Management and Re-delivery of . . . sheep.

Claims on the landlord were frequently made by the farmer for improvement of the stock. On the other hand there might be a claim by the landlord for depreciation to make the farmer become more efficient at looking after his stock. Vets were not yet in being so the farmer became very good at recognising common complaints and diseases and knowing what to do.

Sheep farming – wintering

The sheep farmer's life is a very busy one, but very rewarding. If you do not enjoy the work it would be best to get out as shepherding is a labour of love. At some of the busiest times you have to work all the hours of daylight and sometimes even in the dark if circumstances need it. To get the hoggs, the first-year sheep, off to a good start we would drive them at the 'back end' of the year to better pasture land such as in the vicinity of Coniston or Ulverston. Sometimes they would have to travel 20 miles or so and they would be followed by a horse and trap which was essential to give a lift to a sheep which had grown tired or lame, occasionally one of the dogs would need similar help. Normally we would arrange to meet the farmer who owned the fields about halfway, and in the spring collect them in similar fashion.

In years when there is much snow and strong winds the sheep are sometimes caught under drifts. If there is likelihood of snow the sheep seem to sense this and go for shelter. Savin (juniper) makes a good natural shelter, but in the past shepherds built additional stone shelters for windbreaks and protection from the worst of the rain and snow. Some of these shelters are simply short walls, others in the form of a cross give protection from all directions; some use natural features like the huge boulder above Side House from which a short curving wall has been built. The Herdwicks are really hardy and have been known to survive under a drift for four weeks or more as long as they could get some air. They will eat moss or even their own wool. They love to eat holly of the smooth-leaved type and also ivy. In the past we would cut branches for them. We left the fell gate open so that they could drop in and as soon as they heard the axe they would come running down the hill. If the sheep come through the winter in good fettle it is a great advantage, but after hard winters when the sheep are in poor condition it is only the shepherd's patience and love of the job which brings them through.

The Herdwicks are not only hardy but stocky and agile, able to climb and jump walls. In some areas of the Lake District the sheep are rather hardier, particularly on the highest mountains of Wasdale, Buttermere, Helvellyn, Bowfell, and the Langdale Pikes. We always endeavoured to hire or buy our rams from a farm in those areas. It is most important to keep the flock of Herdwicks pure. Many farmers have crossed them with Swaledales (we call them 'Swaddles'), and now some are almost pure Swaledale. In my opinion they are not nearly so hardy as the pure-bred Herdwicks. The Swaledales are better mothers than the Herdwicks but you need to think of the long-term use. On the whole Herdwick wool is of poor quality but the finer grades are now used in making tweeds and knitted garments. The coarser wool is used for making very good carpets.

To my mind there is nothing better than a bit of Herdwick mutton. All the old butchers and farmers know from experience that it is the sweetest meat you can get. The wethers were always kept until they had been clipped three or four times – then they were more mature and tasted much better. A butcher in Grasmere, called Tom Chew, would not have them until they were this age. He said that they were not mature until then.

The rams are loosed to the ewes about the end of November so that they are lambing at the end of April. Some of the younger farmers now seem to loose a bit earlier but if it is a hard winter this is a disadvantage. The modern method of carrying mineral blocks onto the fells by helicopter is, no doubt, a big help to see the sheep through the winter. Before that the shepherd used to carry hay onto the fells – this was good but it had to be a full three-month meadow hay and of good quality, or the sheep would not eat it and even then they wasted quite a lot. On the farm the old hoggus with a rack round the walls was used for the younger and poorer sheep. But the fell sheep mostly come in quite fit and if they are strong and can stop on the fells they are much better. They are a very independent breed of sheep.

Lambing

Lambing time is one of the busiest times on the farm when you tend the sheep and lambs all the hours of daylight. At the height of lambing you will get as many as 30-40 born each day; some you have to help to lamb. Some of them are coming 'wrong end before', or with a leg back or some such condition: if you were not there on time you would find that the lamb was dead. Herdwicks do not produce many twins. Once we had one with five lambs but they were so small not many of them survived. A sheep cannot maintain twin lambs on these fells – or only for a short period at the end of the season, and not long enough to get the lambs 'heafed' (settled on their native pasture).

If a sheep has a dead lamb, you take a twin off another to put with this ewe. This is done by skinning the dead lamb, putting the skin onto the live one, and rubbing the legs and head with the liver to make it smell the same as her own lamb. Some take to it quite easily and you are able to take the skin off in two or three days – others can be stupid! Some ewes will not even let their own lambs suckle, so these have to be bottle-fed, the same method is used if the ewe does not have enough milk. There was a new product that came onto the market to help the ewe to take to a new lamb; in this case you sprayed the lamb and the sheep's nose, with a specially scented liquid, but I was never very successful with this.

Carrion crows can be a hazard when a lamb has been dropped and has not yet got 'on foot', – they take the tongue out and also go in for the liver. I have seen many lambs walking about without their tongues; they never do any good because they are unable to suckle, or only very poorly. Nature can be very cruel. Sometimes if the crows become very numerous the farmers get together and have a shoot. Foxes can be bad too. I have seen many lambs bitten in the head and still alive, but they never thrive very well. Stray and uncontrolled dogs cause a lot of trouble at lambing time and earlier, when the ewe is carrying.

The lambs are able to stand about 10 minutes after being born, and can run quite sprightly in an hour or so. A nice spell of fine weather at lambing time is a great benefit. If it is wet the lambs never seem able to get dry nor do they manage to get on foot; consequently some will perish.

At the back end of the year farmers liked to collect the rams in before the ewes were in season, then release them on a specific date so you would know when the lambs were due. Occasionally one farmer would not bother and this caused havoc – there would be white (unmarked) lambs born at all times and all over the fells. Sometimes you might find four or five white lambs and they would be wild because they had never been 'dogged' – that is accustomed to the farmer going through them with dogs, gathering them into the fold, marking or cleaning them – they were just wild. White lambs from Pye Howe, Robinson Place, and Harry Place would be all mixed up on the fells and they would tire the dogs out as you tried to gather them.

Many a time I went up the fell on the Pye Howe sled gate, down by Swine Scar Hawse Ghyll into 'Blin Trun Moss' (Blind Tarn Moss) gathering sheep. At the top of the rise, up on the left above the path, is a tiny stone hut, 3 feet wide and 6 feet long, beautifully constructed using slate from the crag. The roof has large slate battens, and the northern wall is formed by a huge piece of bedrock. The entrance is only 2 feet by 2 feet, and the hut is about 2 feet in height. It is not watertight but I have sheltered in there many times when the weather turned bad when I was gathering sheep. This was a difficult job there, the sheep would be hidden amongst the bracken and the savin and among the many rocky knolls – they were very bad to manage as there were little routes between the hillocks and some sheep just don't like to be captured.

Herdwicks living up on the fells tend to have a rather wild nature, not like the more docile field sheep, the Swaddles, Wensleydales, and Suffolks. I well remember gathering sheep by Easedale Tarn one day. I had already gathered a little drove when I saw one odd ewe up on the fellside above the tarn. I sent the dog round behind her to collect her. To my surprise she ran down the hillside and dived straight into the water. She swam out into the middle, then round and round, and eventually just sank without trace.

Lambing time lasts about three weeks, but there are always late arrivals to be seen to. It is quite a worrying time, but if all goes well the farmer will get a good reward and a welcome rest when it is finished.

Clipping and 'clouting'

Clipping is a most important time. The fell sheep are not ready to clip until the end of June or early July. You cannot clip sheep when they have no 'rise'; you cannot clip into the old wool. In the spring, when the sheep start to thrive, the new wool starts to grow and this is what we call the rise. You need to have at least an inch or so before they are easy to clip. If you clip them when they are short of rise it takes a long while and when finished they are bare and like pigs. Then if it comes on a wild wet cold time, which it can, the sheep are cold and sometimes perish. Some stocks of sheep are ready to be clipped sooner than others. The gelt sheep, the ones that have not lambed, were always clipped first. These would include the hoggs and the shearlings which many reckon get off to a better start in life if they do not have a lamb at this early stage.

At a Chapel Stile rummage sale a visitor was puzzled to see the local farmers buying up old tweed suits and coats – he was even more confused when he was told that they were for 'clouting twinters'! If the shearlings or twinters are not very big, as sometimes happens if they have not wintered very well or the summer has not been good, it is not wise to put them with the tup (ram). A method used round the Lake District is to 'clout' them. This is done by sewing a piece of cloth onto the top, and halfway down the tail, to frustrate the tup's approaches. If done properly they can look quite tidy and although not a very quick process, when you get the hang of it, and with a batch of clouts made at the ready, it takes five to ten minutes to do one. If you have a good salving house or a barn it can be done under cover or on a dark wet day. This is done in November when all the sheep are coming into season. If you have a big allotment or intack with a good fence, you can put them into there and, providing no straying rams can get in, do not need to clout them.

We never liked to clip the ewes with lambs before the second week in July. The old-fashioned 'clippings' which took place in the valley have largely died out. At these the farmers would all combine and clip the whole lot on one farm in a day, and then at night they had a dance and sing-song in the barn, a 'merry neet', with plenty to drink. Beer was the main drink and there was ample food. At the clipping the womenfolk would wrap (fold up) the fleeces, carry the beer round and also the tar pot to put on any cuts on the sheep.

The sheep were all clipped by the hand shears and the men sat on a 'sheep creel'. Now it is done by electric clippers and on the ground. Some good clippers can shear a sheep in two minutes whereas clipping with the shears took much longer. In the slump years of the late 1920s the price of wool fell as low as 2¼d. a pound so that it was not really worth the time and labour of clipping the sheep, but it was done for their well-being and health as, when they had a lot of wool on, particularly about the tail, they got very dirty and became infested by maggots. In those days the bluebottle fly was a real menace as they laid their eggs in any dirt on the animal. When the maggots hatched they would literally eat the sheep.

If the weather was humid and damp you had to keep a constant eye on them. This has not been such a big problem since the new dips came out.

Marking

Marking has to be done again after the sheep are clipped. We used to make our own 'smit', a sort of dye, originally with red iron ore collected from deposits on the fells. I remember going with my father to get some of this up White Ghyll. We carried it down on our backs, then it was mixed with Stockholm tar and grease to make the smit for marking. Later on we used to buy a barrel of red powder and mix that in the same way. This was very durable. Nowadays you buy the marking fluid and it rubs off very quickly. The Wool Board are now very fussy about having too much marking in which case they dock the price. Marking the lambs at two or three weeks old, before turning them to the fell, is quite a big job as they all have to be ear-marked as well. Nearly every farm has a different ear mark. Some of the names of the ear marks are:

Slit, split or ritted
Cropped and under bit
Square fork or key bit
Cropped and stump slit
Upper halved, or fore half
Under half, or back half
Cropped
Forked
Cropped and punch holed
Punch holed
Square holed

The following are less commonly used:

Upper shear bit, or fore shear bit
Under bit, shear bit, or back bit
Under key bit, or back key bit
Upper bit, or fore bit
Dove tailed bit
Club ace bit
Spoon shanked
Under slit
Diamond punch holed
Oval punch holed

As can be seen, with two ears, it is possible to get many permutations, such as as 'cropped near ear and upper key bit far ear' or vice versa. The mark we used at Robinson Place was 'forked at both ears'. The *Shepherd's Guides* give all the stocks in the Lake District with both the ear marks and the smit marks. The smit marks are:

Strokes down mid-rib either side
Stroke down the shoulder
Stroke in the lisk (side)
Various pops on fillets (mid-back), rump, or shoulder
Tail head crupping (a stroke down each side of the tail)

With these smit marks, plus the ear marks, there is little danger of many coinciding. There used to be face burns but they were not very good, also big letters marked with boiling pitch, not so much used now. The plastic ear tags being used today do tend to be torn out.

Although the sheep all go on the open fell they mostly do not stray far off their natural 'heaf'. If a sheep goes on Raw Pike and has a lamb there, the lamb will always go onto Raw Pike in the future. There are always the odd one or two which stray and get off their bearings so that shepherds' meetings are held at various places where a shepherd can claim his sheep. Sometimes these strays are sheep which have been sold and have travelled trying to find their old home.

In some areas of the Lake District the common land was 'stinted'. This was so in Borrowdale where a stint was 10 sheep. If a farmer was allowed 50 stints he could keep 500 sheep on the common. Cows, horses, and geese could also be put on the common. We in Langdale were not stinted. When the Commons Registration Act came into force in 1965 every farmer had to register the sheep he was allowed to run on the common. This depended largely on the amount of 'inland' which was the land owned or rented by the farmer. For instance, if a farmer had 26 acres he was allowed about 100 sheep on the common. The registration period lasted for five years after which claims were not accepted.

Salving and dipping

Before dipping was introduced the sheep were washed. Many farmers had their own washfold which might be on a mountain stream where they had some walls built to hold the sheep. They would dam the beck for this purpose. There was a public washfold on the Langdale Beck just below the Grassings (now Wainwrights Inn) in Chapel Stile. There is also a good one at Easedale Tarn where they would push the sheep into the tarn. They had a boat to rescue or turn any back from going into the middle.

Salving was a great yearly ceremony performed for the health of the sheep. This was always done in the autumn and was for the prevention of scab, it also made the fleece waterproof for the winter. This was a slow and tedious job, and one man could salve only about 20 sheep in one day and would have to work by candlelight as the days were so short. The salving dish containing the grease would be pushed into the wall at a convenient height, the farmer sat with the sheep on the sheep creel. He parted the wool and then applied the salve, rubbing it in with his fingers. This was repeated all over the skin of the animal. It was an expensive job since the salve was made of tar and butter. In 1646 a Langdale farmer paid these sums for salve and labour:

6 stone of butter	£5 – 3 – 0
5 men	£1 – 7 – 0

8 of tallow		£2 – 13 – 4
3½ loads of tar		£4 – 10 – 0
	Total	£13 – 13 – 4

I don't think the tallow was always used. In those days the expense was very heavy. In some cases salvers were hired and payment was 6d. per day in 1682, quite a lot of money in those days. In lots of cases ale was supplied, it was 3d. a gallon in 1678. Salving was dispensed with towards the end of the 19th century when dipping was introduced.

The first dipping tubs were made of wood and the summer dips were mainly for protection from maggots, scab, and ticks by use of arsenical compounds, but these soon washed out in wet weather. Now the summer dips are made of DDT and Gamaxene types, discovered during the 1939 war. The 'back end' (autumn) dips are more of a greasy nature. This was for scab and also to make the fleece waterproof. A few years ago there were two compulsory dipping periods and notice had to be given to the police who often came to watch to see that the work was done properly, though a good farmer would always dip his sheep thoroughly without needing to be watched. There was a period, a few years ago, when these two dippings were relaxed. This was found to be a big mistake with sheep scab breaking out in various places, so compulsory dipping was again introduced.

A lamb, after dipping in the autumn or at Michaelmas, becomes known as a 'hogg'; when it gets its first two broad teeth, as well as four small teeth, it becomes a 'twinter' or 'two shear'. The next year it casts two small teeth which are replaced by two more broad teeth, thus having four broad teeth, and becomes a 'thrinter' or 'three shear'. Next year when it casts the remaining two small teeth and gets a full mouth of six broad teeth it becomes a ewe, usually at four years. When a lamb is castrated (or gelt) it is a wether. An animal that is a weakling and will not thrive is called a 'reckling' or 'runt'.

It became the fashion, if you had some of the older ewes and some that were not up to the good standard of breeding Herdwicks, to put them to a ram of some finer breed such as a Suffolk, Border Leicester, Wensleydale, or the like, to produce a few half-bred lambs. But these had to be kept off the fells, on better land. It was no use expecting them to do well on these fells.

Sheepdogs

One of the most important animals on the farm is the sheepdog – some of these dogs have more sense than human beings! They are worth a great deal of money. Most farmers would have four or five to ensure that enough were fully active as some would go lame or ill from time to time or simply get old. They are also such good companions. You never feel lonely on the fell even on the bleakest day. The fact is that you just cannot do without them. I think the sheepdog is the most intelligent of all the breeds of dogs.

I always prefered a bitch or female as they seemed to me to be more faithful. However, I must mention one male dog I had called Don. He was a grey, shaggy dog with quite a distinct 'wall eye' (one eye of a different colour) – one was very blue indeed. He was a good dog and very faithful and lived to be quite a good age, about 10 years. When any of my dogs died or had to be put to sleep I always shed a few tears. If a good faithful dog of mine had to be put to sleep it was always a hard decision to make.

The bitches I had were all very faithful; one called Lady was jet black and had some outstanding qualities. There are two types of dogs: some are good drivers and some are good fetchers – she was a good all-round dog. She had one bad habit – she loved chasing cars, and liked waiting for them to come! Another I had was Lyn; she was not the usual black, but a real ginger brown. She was slim and fast, not many sheep would get away from her. I also had two bitches which were great pals, one called Bute was quite small and rather like a spaniel, not too brilliant at work but very faithful. If I put my jacket down while working anywhere, she would lie on it and would not let anyone else take it. The other was a brownish colour and I called her Lady and gave her the nickname Willie Woof. This was the most outstanding of all my dogs. She also had a wall eye but she could get sheep out from places which no other dog could manage to. She knew whatever I said and I am sure that she had more sense than a lot of human beings. If I had her in the car with me she would not let anybody in the car unless I was there. She loved children and as soon as she saw a tiny tot she would go and make a fuss and lick their face. If she was left outside, she would knock at the door. We had a very big knocker on our back door and if you watched you would see her lift the knocker with her nose. This dog was quite outstanding. I have a photograph of her which I treasure.

All these dogs were really marvellous and I am sure that I could not have managed without them. A good sheepdog is certainly worth its weight in gold. The average sheepdog lives to about 10 or 12 years, and although they cannot stay the pace after 8 or 9 they still love to go. I think they want to go to see if you are doing your job right!

Diseases – treatments and controls

Sheep are not very prone to disease and Herdwicks are particularly hardy. Most diseases can now be controlled by injection with vaccines and serums, mainly for protection. Some of these are 'six-in-one' and can control things like dysentry, pulpy kidney, blackleg, braxy, and tetanus. Things like fluke and worms can be cured by using a drench (liquid medicine). The fluke is caused by a snail and is mostly picked up off grass in marshy places; they make their way to the liver and destroy it if not caught in time. Another complaint is sturdy, this is a tumour on the brain and is caused by an egg that has been passed through a dog, so it is necessary to keep your dog wormed regularly. To cure this you had to bore a hole in the scalp with a red-hot iron called a 'sturdy iron'. If you got the right place it came out quite easily, it is about the size of an egg. Afterwards you put a patch of pitch, which had been melted first. Fortunately this is not very common. It is surprising what you can do if things are caught in time and you have a bit of knowledge.

Sheep scab and maggots are now under control provided the sheep are dipped thoroughly. Penicillin is used now as an injection if septicaemia is badly set in. Moss illness or 'drop' is a lambing-time ailment and is a form of milk fever. You have to be on constant guard at this time of the year as the ewes just drop down without any symptoms of anything being wrong. If they drop in an out-of-the way place where they cannot be seen, it is fatal. Sometimes they look as if they are dead but if injected with 80-90ccs of calcium borogluconate under the skin they are often standing on their feet before you have finished the job. If you are wise you carry the bottle and the syringe with you. Some stocks of sheep are more prone to this than others and some years are worse than others. In some cases, if you put the sheep somewhere where they are discontented or unhappy, even just to a fresh pasture, drop can occur. There seems to be an element of sulk attached to it as well as a shortage of calcium.

Another rather nasty complaint struck out 20-30 years ago. This is called 'scrapie'. It is a virus that destroys the cells of the brain and up to now there is not much known as to the cure or the prevention. A friend of mine, working on this at the experimental station in Edinburgh, maintained it had never occurred in Herdwick sheep, so he invited me to go and see the various animals with it and then I could identify it in the Herdwicks if they became infected. I went up to Edinburgh on 15 September 1961. It was most interesting to see the various breeds of sheep with the disease, but it was also very pathetic. A few years later, I had one sheep which to me seemed to show the symptoms so I rang him and he came down to collect it. They had to slaughter it and take the brain out. He rang me later to say that it had been identified as scrapie, so his earlier views had to be modified.

'The Shepherd'

A real Herdwick sheep man has not a lot of time for cows and prefers to be working and tending his sheep. My brother Fletcher was one of those. Indeed he knew all his sheep and what part of the fell they frequented. He would rather go to the fell than look after the cows. So this had to be my job. I am sure he did not like milking as he invariably found some excuse to go amongst his beloved sheep. The Revd Herbert Bell wrote a poem while walking over the fell from Grasmere on his 90th birthday – 20 September 1979 – it was called 'The Shepherd' and he said it was inspired by seeing my brother and was written as an epitaph to him:

A form, a person, standing at ease,
Leaning on a shepherd's crook.
High above the world of human noises.
No speech, only a silent stillness.
The air breathes gently,
Rustling over the waves and rocky pinnacles.
He stood, regarding while our young pastime
Played over his moors, his home.
The vital task of watching.
He knew his sheep;
His eye and mind knew every one
In storm and in motherhood.
We were made strong by watching him.
When the storm takes the shepherd and the sheep
New life is ever playing,
New trials and new glories.

Counting sheep

The shepherd's method of counting sheep varies from valley to valley, some have the same numbers, other are a mixture and some have some unique features. In Langdale we would count them as follows:

1. Yan
2. Taen
3. Tethera
4. Methera
5. Mint (or Pimp)
6. Pipe (or Sethera)
7. Platus
8. Overa
9. Dovera
10. Dick
11. Yan a dick
12. Taen a dick
13. Tethera dick
14. Methera dick
15. Gigot
16. Yana gigot
17. Taena gigot
18. Tethera gigot
19. Methera gigot
20. Bumfit

When you reached 20 you put a pebble in your pocket. So that if you had 10 pebbles in your pocket you would have counted 200 sheep. This method, close to the Borrowdale way of counting, is supposed to have originated from the Norsemens' language where much of our dialect has come from.

Sheep-stealing

A Langdale farmer was accused of stealing sheep early in this century. The farmers in the valley had been losing sheep and had become suspicious of this man. He was tried at Appleby Assizes and was acquitted, although he was caught with one of Mr Buntin's sheep, ready to kill it, and he had four sheepskins at a fellmonger's with the marks of T.H. Youdell of Millbeck. He pleaded that he thought that they were his own sheep and was acquitted, as the jury did not consider there was enough evidence to convict him. He came back to Langdale and continued to kill sheep and take the meat to his customers. Then he started to kill his own stock sheep, since times were hard and he had a family of five to support. He eventually went into liquidation, and out of a fell stock of 350 sheep which he had taken over, only 136 remained. In years gone by sheep-stealing was a hanging offence. Rather rough justice, I think.

Lost sheep

Sheep are prone to wander and some are worse than others. From time to time some would wander over from Grasmere or Borrowdale or Eskdale into Langdale and vice versa. By the ear or smit marks you could tell whose they were. You would care for these strays, sometimes for three weeks, sometimes for three months, making sure that they were well fed until it was time to return them. Exchanging the lost sheep used to be done by meeting the other shepherds halfway. The Grasmere shepherds would meet the Langdale chaps at the top of the Pye Howe sled road. The time was stated and if they were not there you waited – if you went over the top you would always have to go the extra trek! Keith Rowand remembers going to meet the Borrowdale shepherds at the top of Stake Pass – it took him two hours to get there with his sheep, the men from Borrowdale had further to travel and it took them three hours. The top of Wrynose was the meeting place with the Eskdale shepherds. This took place at lambing time, clipping time, and at the back end. Now it is all done by Landrover and trailer – you see little boxes of sheep dashing along the roads in the late evening or early morning to avoid the worst of the traffic – it is a much quicker method. The sheep which were difficult to identify would all go to the various shepherds' meets around the Lake District.

CHAPTER 5

FARMING LIFE

Hedging

The old hedges were planted on a bank of soil to make a high barrier to keep the stock in. They needed attention from time to time and were laid every 5-7 years depending on their condition. We would say 'We're going to lig (lay) the hedge today'. To do this the old dead wood was cut or pulled out and then posts were hammered in about 2 feet 6 inches apart. These posts were made from ash, hazel, or whatever was at hand. Any wood of 6-8 inches diameter would be riven into three or four pieces. The boughs of the young trees in the hedge were then nicked almost through but leaving a little for the sap to run. They were laid (bent down) always to the north to prevent the wind causing havoc, and then held in place by 'yedders', long strips of willow, hazel, or ash, using the straightest pieces available. The yedders were woven between the posts as in basket-making, interlocking with each other. About 50 yedders might be needed from one wall to a gate. To give the hedge a good finish you would knock the posts in so that they were all level, if you hit a stone you would saw a piece off the top. If a wind got up before yeddering was completed branches might blow up and break off, so if we hadn't finished by night time we always put a boulder or big tree trunk on to prevent this happening. Willows make the best hedges, hazel was good, ash all right – thorn was difficult to work although it made a good hedge.

Cattle

All the farms had a small herd of cattle about 20 years ago. The chief breed was the Shorthorns, then the Friesians seemed to take over, they were good milkers. I like the Shorthorns; the roan and red ones were really good to look at. They all had nice horns, but now all the cattle are dehorned. We had about 30 cattle, milking about eight during the summer. The cows were sold after they had calved a couple of times, and were worth a lot then. Now many farmers let the calves suckle the cows. We used to milk the cows and bucket-feed the calves. When the calves had reached a certain age they were fed with skimmed milk plus some meal or gruel. The sucklers take up less time and labour which is an advantage in these days of high wages. Now there are various breeds of cattle in Langdale – Angus, Galloway, and some of the newer breeds that have come over from France.

In the winter when there was a hard frost after we had finished milking, many a time we would go up from Robinson Place to Youdell Tarn skating in the moonlight; it was a bit rough but great fun.

Haymaking

In years gone by hay was made by hand and in a bad summer it was a long-drawn-out business. It all had to be scaled and turned and cocked, and when ready it was forked onto the carts loose. Then came along all the modern implements and nowadays it is all baled, which is much quicker, and you can lead a 10-acre field in a very short time. A good deal of grass is made into silage and either put into a pit or into plastic bags. This is all a big saving of time, but I like the old fashioned haymaking best. There was nothing more pleasant than the smell of new-mown hay. Times seem to change so quickly, but I don't think you get the same joy as we used to. Having your tea under a tree in the hayfield was really something to appreciate.

It was lovely to see a row of people turning and scaling hay, and forking it into the horse-drawn carts; it was a delight to see the women coming into the field at 3.30pm with a basket of home-made fare and a large can of tea. Any tea left to go cold was a real pleasure to wash the dust away. Mother used to make a quantity of 'botanic beer' which was brought into the fields and left behind the hedge for the following day. It was all so enjoyable and I for one enjoyed my haymaking days. It was hard work especially if the weather was 'catchy'. It was a great joy to see the last load of hay being taken in. Often in a wet summer we were haymaking in late September, and occasionally into October. There was no baling then. It was all led loose and put into the hay 'moo' in the barn. When it was being fed to the animals it was pulled out by hand using a 'hay puller' and shaken up thoroughly to get all the dust out – if it had been a bad summer and hard to dry it would be very dusty. For the Herdwicks it had to be the very best hay!

Horses, pigs, and poultry

All the farms had two horses which were kept for doing the heavy farm work and nearly all used to keep a couple of pigs and kill and cure them for their own use. I miss the home-cured bacon – if it was cured properly it was really very tasty. Everybody kept a few hens and hatched all their own chickens. We always put 13 eggs under the broody hen and when they were hatching they were marvellous to see. You could help the chick out by sprinkling the egg with water. Sometimes a hen would lay away outside and sit on them by herself; it was nice to see her turn out with a clutch of chicks. She had to be lucky to escape the fox. The eggs from the free-range hens are much better than those from battery hens and we could get a better price for them. We used to put sweet gale or bog myrtle, *myrica gale*, in the henhouse to counteract lice.

A few ducks were kept too, as well as geese and turkeys, but the turkeys were a lot of work as you lost a good number. But now with all the modern medicines they are much easier to rear. However, not much poultry is kept nowadays.

Flooding – 1937

If it rains very heavily, Great Langdale can have severe flooding as the valley is so flat. This was particularly bad before the Drainage Board came to work on the main beck in 1942. They dredged it and banked up the sides and it has not been as bad since, although it can still be a hazard. Before this was done it was necessary to bring the livestock up onto higher ground at night if flooding was likely to happen.

One particular case happened in 1937, on 5 September to be precise. The Greenhows of Harry Place left their two horses too long in the big pasture opposite Robinson Place. This field is always subject to severe flooding. The horses were standing in the middle of the field on slightly higher ground and it was not possible to get them from our side and they would not come for shouting or anything else. This was now being witnessed by quite a crowd of people. The flood water grew higher and higher, first up to their bellies and then all you could see was their heads. My brother Fletcher and Harry Greenhow got onto our horse and went round the back side of them to try to coax them that way but without success. Eventually a fellow who was camping near by got onto an inflatable bed and went out to the other side of them. As soon as they saw him, they were very frightened and dived in and swam out to us. They didn't seem any worse for their experience. It is marvellous how these animals could swim as no doubt they had never swum before.

Worst storm in living memory – 1966

The Langdales were hit by a horrifying storm on the the night of Saturday, 13 August 1966. For six hours the valleys were lit up with nearly 200 flashes of lightening, many striking crags. The Preston Meteorological Centre reported that the thunder clouds were towering to at least six miles and the centre was over the Langdale Pikes where it was estimated that 4-5 inches of rain per hour must have been falling. The *Westmorland Gazette* for 19 August carried a full report:

Night of Terror as Huge Boulders crash down Fellsides

Devastation follows Langdale Deluge

A violent thunderstorm following 12 hours of incessant rain produced a night of terror in the Langdale valley . . . on Saturday. Half-a-dozen mountain streams, swollen to fierce torrents, burst their banks, sweeping away stone walls and parts of the road. The flood waters raged through two hotels, farms and several houses and inundated the National Trust camping site at the head of the valley. The onrush tore away parts of the fellsides and turned Elterwater and the northern reaches of Lake Windermere brown. As they receded on Sunday, the scene at the head of the valley was of devastation.

A car, swept 40 yards off the road, was standing with its doors wide open, drying out in the middle of a field; a local farmer had been to collect his two cows, miraculously little the worse after being swept four miles down Langdale Beck to the village of Chapel Stile.

At the New Dungeon Ghyll Hotel, the proprietor, Mr. Cyril G. Bulman, assessed his losses at more than £1,000. The waters of Mill Ghyll swept in at the rear of the hotel at 9.30pm and out of the front door, flooding the entire ground floor to a depth of 3 ft. 'There were about 30 people in the hotel at the time when the water began to come in', a student worker at the hotel commented the same night, 'and in a matter of minutes it was like a swimming pool'. The road leading from the hotel was completely flooded and a dozen motorists stayed the night in their cars.

'Some of the boulders which careered down the hillside and finished up in my back garden must weigh half a ton,' Mr. Bulman said. 'The noise they made rumbling down the fellside was terrifying. I have never known flooding like this and I have lived in the valley for 30 years'. The bar was closed for two days while mopping up operations were carried out. The twenty guests at the hotel insisted on staying on and most of them were out walking on the fells the next day.

Pastures which were smothered in mud will be almost useless for grazing this year. A road bridge was swept away when Rossett Ghyll, another mountain stream, overflowed. Stool End Farm, occupied by Mr. Harry Cowman, was completely cut off from the rest of the valley, but the collapse of the bridge probably prevented another farm from being swamped.

Mr. Hugh Parker of Middlefell Farm, which escaped the flooding, told the Gazette how three of his cows had escaped injury after being swept away. 'I recovered one which had been carried half a mile, and the other two finished up at Bayesbrown Farm, four miles away at Chapel Stile. They were knocked about and skinned, but as far as I can see they are alright otherwise'.

Huge quantities of stones and gravel covered part of the road which has a 1 in 4 gradient leading to Blea Tarn, which was also undermined and is likely to be closed to traffic for some time. The route from Little Langdale to the summit of Wrynose Pass was also impassable.

Household chores

I must mention the woman of the house and the tasks she had to perform – she had, more or less, something planned for every day of the week.

Monday was wash-day – a big task in days gone by. First the washing boiler had to be filled with water, some held at least 20-30 gallons. The fire underneath was usually stoked with wood or sticks which had to be collected beforehand. We, as children, had to gather kindling and wood and we would help to use the old 'dollylegs'; some used the 'posser'. The clothes were all put in the dolly tub with water and it was quite an art to use the dollylegs effectively – this job got a lot of the dirt out; then they were rubbed and scrubbed out in the wash-house. When more or less clean they were put into the boiler and boiled. But not the woollens! You had to be careful with these as they would shrink. The clothes were then mangled – we used to help to turn the old mangle – and then they were all hung out to dry. This could be a problem if the weather was not good.

After they were dry they were neatly ironed. My mother usually used a charcoal iron. We bought the charcoal and put it in the fire until it was red hot. The iron had a lid which you lifted to put the hot charcoal in, and a hole in the back for the air, and a chimney. You had to keep removing the old bits of cold charcoal. She sometimes used a flat iron which had to be reheated frequently, so you would use two at a time, one being heated while you ironed with the other. She also had a box iron – this was deep, a shell with a handle and two pieces of iron, one piece you put into the fire until red hot and then slipped into a sliding door, it was then ready for use. Some of the items had to be starched. When finished they were all hung on the clothes-horse before being put away. Some of the garments had to be darned, mended, or patched – this was mostly done in the evenings, and in a household with a large family it was a big job. Often there was a clothes-rack hanging by pulleys from the ceiling where they could be aired – in some cases to be dried if the weather was bad.

Baking-day too was quite a big task. My mother always bought 10 stone of white flour every month and possibly some brown as well – this usually lasted the month. She made all her own bread, pasties, and cakes. I can still see her kneading the bread in the large bowl or dish. On its first rise it was put down, and after its second rise it was put in the bread tins and set to rise again in front of the fire. The old-fashioned oven had the boiler at one side and the oven at the other. The 'nigget' (a movable iron plate) was taken out so that the heat could circulate round the oven. This had to be heated first with a mixture of coal and wood. One day my mother was using the word 'nigget' and my sister said 'There's no such word', I retorted 'Oh yes there is – it's in the dictionary'. I got the dictionary out and looked it up. It wasn't there but, making it all up, I said 'Here it is – nigget – a device for keeping the fire from the oven when not in use'!

I can still remember the lovely smell of new-baked bread. At Christmas time currants and raisins were added, together with chopped candied orange and lemon – she made what we called currant bread – this was really quite a delicacy as not many cakes were made for money was not so plentiful. My mother used to tell of going to a tea-party where there was nothing else but white bread and jam plus the currant bread instead of cakes. Rum butter was a good old-fashioned addition which was greatly enjoyed – a big bowl of rum butter was always made for Christmas and also for christenings.

Bath-night was quite a job with a big family. The old wooden tub or zinc bath was placed in front of the fire and the hot water was provided by the fireside boiler which was kept filled with cold water using a bucket. Some of these boilers had a brass tap at the front where you could draw your water; others had a large lid at the top and you had to ladle it out from the top as they had no tap. In some ways these were better because taps became leaky and then you had to stand a jam jar under to catch the drops.

The fireside was always kept nice and clean and once a week the grate would be black-leaded and any brass pieces polished. The fender at the front was always bright – it was cleaned with bath brick and 'Shinio'. The fire irons were also polished and kept bright. There is nothing more beautiful than a well-cared-for fireside, but it was all done by a great amount of hard work.

The old stone-flagged rooms were all scrubbed once a week, more if need arose. Some of the flags were of the green type but they were not as smooth as the blue Brathay ones. They all looked so attractive with the various rugs and hand-pegged hearth rugs which were all made by the women of the house. They took a great deal of pride in the pegging of these rugs to make some lovely patterns.

Oilcloth was used on a lot of kitchen tables but on special occasions a white embroidered cloth was used. Crocheting and needlework were worked a lot in those days and I can never remember my mother sitting doing nothing! She was always knitting, tatting, crocheting, or mending or darning something. She always went out to help to milk as my father had accidentally cut off his thumb and was not much good at the job. She always fed the calves, the poultry, and the pigs.

The old-fashioned long brush was used to sweep the floors and the hand brush and small shovel was used in awkward places such as under the beds. Sometimes if there was a lot of dust my mother would sprinkle moist tea leaves to damp it down before brushing it up. Before stainless steel came in the knives all had to be cleaned – we had a knife board, a piece of wood with a piece of leather fixed in. You sprinkled or scraped some bath brick on the board and rubbed until the knives were clean; this job was often given to us as children to do. It was quite a laborious task and we soon got fed up with it. At about 12 years of age we would have to help muck out and milk the cows before going to school. We all wore clogs with caulkers (irons) on the bottom and these had to be cleaned before going to school – no wellingtons in those days.

During the First War we made our own tallow candles. We killed our own sheep and the spare mutton fat which was no good for anything else was heated up in a large pan. We would hold a piece of string or wick and lower it into the heated mutton fat and keep dipping it in until it got the required thickness. You had to hold the string taut when you lifted it out. The other method was using a wick in a metal tube into which you poured the heated fat and released it from the mould when cold. It burned with a very dirty flame and a little bit of a smell, it always ran down the edge and made a mess. When we were young we sometimes put tiny spots of candle fat on the back of our hand and then showed it to mother – 'Look – I've got measles'! We would also make our own soap by boiling up the mutton fat with caustic soda.

Sunday was always observed as a day of rest and we were not allowed to do any sort of needlework or to play cards. If it was a fine day mother would take us for a walk, to gather some flowers or look for birds' nests – but we had not to touch the nest as she said that the birds would forsake it. My father played the violin and sang while mother played the piano and we would all sing songs or hymns. I didn't go to Sunday School as it was so far – we had walked to school all through the week, I would go to church in the evenings when Hannah was playing the organ. We really enjoyed our young life and we were never bored.

Old farm tools and methods

Muck drag or dung drag – used to pull the manure out of the cart to be left in heaps a certain distance apart. The drag was also used for cleaning weeds from streams.

Peat-cutting spade – a flat heart-shaped blade for cutting peat.

Draining spade – a narrow spade for digging drains.

Hay rake – essential in the hay field. The 'man rake' or 'rover' was 4 feet long with curved metal teeth and used to rake up after loading the cart.

Scythe or ley – for cutting grass in awkward places such as the edges of fields, and for mowing bracken and other weeds.

Sickle – for cutting briars and the like in awkward places.

Turnip knife – for snagging turnips. This had a spike on the end to grab the turnip and was a very dangerous tool.

Hay knife – when the hay was carted loose and packed in the hay mow (pronounced 'moo') in the barn the hay knife was used to cut a layer down on one side of the mow.

Hay puller – used for pulling the hay out of the mow. Some people preferred to do it like this but it was a slow process.

Flail – for threshing the corn, a long handle with another piece of wood fixed on the end by two pieces of leather.

Mattock – had a broad blade and used when you were grubbing at a tree.

Pick – a sharp pointed tool for digging, essential for extracting sammel.

Billhook or back bill – for fencing and chopping trees back, sharp on both sides.

Fleam – a tool for bleeding animals like a knife with three or four different sized blades. You bled the animal at the roof of the mouth.

Blocking stick – used for killing animals like cows – it was held on the animal's forehead and hit with a hammer.

Pestle and mortar – for grinding things to powder when dosing animals.

Drenching horn – this was for dosing an animal with medicine and was made from a cow's horn. Pigs were drenched with an old boot with the toe cut out.

Butter churn – for churning the cream to make butter.

Salve pan – pronounced 'sove' as in Hove; this was made from a piece of wood hewn out like a ladle with a piece of wood nailed on the bottom to stick in the wall near where you were doing the job.

Sheep creel – a long stool often made quite primitively with two blocks of wood, one at each end and joined with a long piece of wood, and legs. It was used when clipping, cleaning, and salving sheep. A big one was used when killing a pig – this was then called a pig creel.

Shears – for clipping and cleaning sheep.

Crooks – a dipping crook for pushing the sheep under in the dipping tub, a shepherd's crook for catching the sheep by the neck, and a leg crook, made of metal, for catching them by the leg.

Wool hook – for hauling the sheets (hessian bags) of wool about after packing – a full sheet would weigh nearly 2 hundredweight.

Two-hand barrow – a very useful gadget for carrying heavy stones.

Sledge or sled – for carting peat down the hillside, also for sledging bracken for bedding and stones for walling.

Mole trap – the early ones were made of wood, now metal.

Snare – for catching hares.

Rabbit and rat traps – made of steel wire, but now illegal.

Swill – a shallow basket, made of torn strips of wood, preferably oak, for picking up potatoes or sticks for the fire.

Pulley – very useful for lifting heavy objects such as beams onto roofs.

Hoe – for weeding between rows.

Stitch harrow – pulled by a horse down the rows of crops for weeding; a 'stitch' is a row of any crop, potatoes etc.

Brush harrow – made by binding birch branches together for harrowing manure – these are now made of chain.

Fork or gripe – for spreading manure or loading onto the cart.

Snow plough – pulled by horses and was often kept at Wall End, where the farmer cleared the snow from the end of the valley.

Crosscut saw – used to cut up big logs and felling trees – it was an endless job – I have sawed wood for days on end.

Castrating irons – used to castrate bulls and tups. Lambs were drawn by your teeth or with castrating pliers.

Separator – a modern gadget for separating the cream from the milk.

Cheese press – used for pressing the whey from the solids.

Breaking kettle – made of brass or copper and holding about 10 gallons. The milk was put into this and rennet added to make it curdle. It took a lot of milk to make one pound of cheese.

CHAPTER 6

INDUSTRIES, CRAFTS AND COMMERCE

Elterwater Gunpowder Works

The Elterwater Gunpowder Works were founded in 1824 by David Huddleston who had returned to his native place (he was born in Little Langdale) after being a banker in Kendal. After he died in 1831 the works continued in a flourishing way for many years. They were eventually taken over by Nobel Ltd and later by I.C.I. and carried on for quite a long while, being very successful during the First World War. But in the 1920s gunpowder was not required in such large quantities and the works became run down.

After a number of years they found that the site was not big enough, so in 1850 it was necessary for a bit more of the Elterwater Common to be taken in. I have an Ordnance Survey map dated 1861 which shows the old boundary. They took a piece of the common to add to their works and built a good high wall round it and a new piece of road. The area was thickly wooded to stop the blast if there was an explosion. The wheels were driven by water from Great Langdale Beck. A dam had been built opposite Weir Cottages and the water conveyed by means of a sluice. They found that in a dry period there was not sufficient water to drive the machines so they obtained permission to make a dam at Stickle Tarn. When it was dry weather I can still remember seeing a man go up to the dam on a Friday night to stop it up and then go again on Sunday night to open it. In about 1860 the company built the schoolmaster's house at Brig Howe in recognition that the commoners had given permission for part of the common to be enclosed for the company's use. Mr Samuel Twist was the schoolmaster at the time.

In the 1890s the firm employed about 80 people, both men and women. The women were used mostly as packers. The powder had quite a strong smell and, if you met people coming from work, you could smell them before you actually reached them. The local name for the people working there was 'powder monkeys'.

Some parts of the works were very dangerous, the corning house being the worst. There the 'incorporated powder', a mixture of carbon, sulphur, and saltpetre, which had previously been pressed together, was broken up into small grains before being coated with black lead. Everywhere people tried to avoid sparks by using brass hammers and scissors, and the horses wore brass shoes. In spite of all these precautions there were a few explosions, the worst being in September 1916 when four men were killed, Messrs Joiner, Carradus, Atkinson, and Foxcroft. They were all working in the corning house. I was in school at the time and the explosion shook the school; Miss Gaskell, our teacher, said 'That's the corning house'. I was sitting next to Joe Joiner whose father was one of the men killed. The explosion was very bad; it broke a lot of windows in Elterwater village. I can remember my mother taking us across the beck to view what had been the corning house. The roof had been blown off and was resting in the top of a tree. The four men were buried just inside the gate of the new cemetery which had been opened only a short time.

The factory made the black powder but most of the other materials, including sulphur and saltpetre, had to be conveyed from Windermere Station by horse and cart. The company had land and made hay to feed the horses, and their own blacksmith's shop on the opposite side of the Chapel Stile road, which is still called 'Cylinders'. The retorts for obtaining the pure carbon were cylindrical, giving the name to this site.

Every day at least four tons of gunpowder were carted down the valley to be transported away by train from Windermere. This was all used for civil engineering and not in instruments of war. Some was exported to South America and West Africa.

The whole complex closed in 1929. All documents were meticulously destroyed and the buildings were deliberately burnt down to comply with the safety measures. The big trees were cut down and sold for timber. It was a real shambles as one can imagine. I remember getting permission to go in and get posts out of the mass of tangled boughs. My father and brother and I went in to cut out and prepare this useful wood and we got a local man called Jack Elleray with a one-ton lorry to lead it up to Robinson Place. Some of the pieces were thick and we took them to get them sawn up at a saw-shed at Skelwith Bridge which was run by a man called Fred Burns.

The men who were made redundant were transferred to other I.C.I. works, one of which was at Gatebeck. The whole site was sold at very low prices as it was in the slump years. The people who lived in the company's houses all had a chance to buy their houses very cheaply. I.C.I. owned a lot of property round Elterwater and Chapel Stile. The actual factory was bought by R.L. Hall who, with the help of Harry Pearce, built it up and made it into a holiday caravan site, and named it the Langdale Estate. It has recently been taken over and made into a luxury timeshare complex with sports facilities including squash court, swimming pool, and restaurants.

Skelwith Bridge corn mill and bobbin mill

About 1684 Reginald Holme owned a corn mill at Skelwith Bridge. He was a man of turbulent character and was many times in trouble. On one occasion he built a dam or weir across the beck for the good of his mill, but to the damage, it was declared, of the footpath above and of his neighbours' grounds. The Rydal squire twice headed a party for the forcible destruction of the dam. A long time afterwards Holme was in fierce conflict with Michael Satterthwaite of Langdale about this or another dam. As well as a corn mill on this site there was probably a walking mill, for fulling cloth, which took advantage of this fine flow of water.

At a later date Jeremiah Coward started his Skelwith Bridge bobbin mill. He was born in 1789 and lived until he was 82 years old. His father, John, bought the old Hare and Hounds Inn, now the Skelwith Bridge Hotel, for £750. Mr Coward helped to form the local Prosecution Bond at Ambleside, which still exists, for the prosecution of felons. A Prosecution Bond Dinner for men is held each January at which the guest speaker is a well-known personality. The Prosecution Bond was formed on 23 February 1813, coming into existence before the police force.

Vast numbers of bobbins were needed for the Lancashire cotton mills so Lakeland's bobbin works became nationally important. Between 1847 and 1870 49 such works were operating in the area. However, Norwegian competition and the improvement in textile machinery, together with a decline in the cotton industry resulted in the closure of many bobbin mills. Mr Coward's mill had a varied life and at one time was making and supplying box timber for blasting cartridges. Today it is a slate and stone works. It was driven by water power, a sluice conveying the water to the mill from the beck near the Skelwith waterfalls – the remains of the race can still be seen today.

Charcoal-burning

Charcoal was used by the iron industry for smelting. To make the charcoal a huge pile of hazel sticks was carefully arranged, heaped up, and then covered with turf. Hot burning charcoal was introduced at the top, and the pits burned, against nature, from the top downwards. The pit was allowed to smoulder but never flare. A burn could last anything from two, three, or four days. This method left the charcoal intact whereas, if it had been lit from the bottom, the weight of the unburnt branches would have crushed the charcoal. The men slept in their clothes and at times their wives lived in the huts with them.

In the intakes between Harry Place and Pye Howe there are at least half a dozen 'pitsteads' still in good condition. Traces of charcoal and cinders have been found there recently, which indicates that the fell breast was once quite thickly wooded. There are not many trees growing there now – none of the necessary hazel, just a goodly number of hawthorn which make a good show of colour in the spring. None of the pitsteads were high up the fellside as the trees would not grow very well up there. The pitsteads are quite close together; they have been dug out, flattened, and are circular. I suppose the nearer they were to the trees the less distance they had to bring the wood, a necessary consideration when it all had to be carried by hand. Later the charcoal would all have to be brought down the fellside, but this could be done by packhorse or horse sled.

Slate-quarrying

Many folk have at some time or other wandered through the beautiful and mountainous valleys of Lakeland and have noticed the old farmhouses and barns. These buildings seem to remain as relics of years gone by, being no doubt built and roofed by local material hundreds of years ago. Some of them are known to have been built in the 16th century and have withstood all the ravages of time and tempest. It is plainly evident that our yeomen of the fell dales had very definite ideas about building and roofing their houses. In some of the old barns the principal beams have been reared with the bottom end resting on the ground – cruck beams they are called – the walls being built around them afterwards. The beams have been hewn from a tree with an axe and fastened together with wooden pegs.

Our fore-elders undoubtedly acquired their roofing slates from the local slate which abounds in some of the mountainous districts of Cumberland, Westmorland, and Lancashire. 'It would appear that some of the Roman buildings, including the fort at Ambleside and the barrack blocks at Hardknott, were roofed with local slate' (Pearsall and Pennington, 1973). Some of the slates used long ago were of great thickness and were more like flags. In some cases moss was stuffed into the holes to absorb the wet which ran down from the ill-fitting slates, the moss required frequent replenishment. There is no doubt that roofing slates or battens of a sort were quarried from these rocks before gunpowder was used for blasting in England. We know of limestone being used to burst open rocks in early times. The Silurian green slate of the Buttermere, Langdale, and Coniston districts formed an abundant supply of this valuable asset, particularly suitable for roofing purposes.

Early in the last century the green slate was being extensively quarried in these three districts. There were also excellent blue slate quarries in Ulpha, Cockermouth, Buttermere, Borrowdale, and Bassenthwaite. Probably the green slate was in some ways superior as this in due course was prefered to the blue.

By about 1832 quarries were being worked at Thrang Crag in Langdale and at other places in the district. The primitive tools used in olden times have given way to up-to-date working machinery and tools. The old-fashioned method of boring with hammer and jumper has been superseded with the boring-machine driven by the compressor. For the green slate quarries at Buttermere and Langdale a fine up-to-date plant has been installed to deal efficiently with every phase of the working. Saws cut the quarried rock into the correct sizes ready to be sent into the 'rivers'. A few years ago machines for dressing the slate were tried, but it was found that the old system of hand-dressing was more economical and better in every way. The tendency to roof buildings with other, cheaper materials has increased lately but, in my opinion, nothing can be as near, or as much in keeping with, the natural beauty of the Lake District as a house built with local stone and roofed with the beautiful green slate which has been produced in the district by local labour.

Langdale slate and stone is very popular in countries as far away as America and is exported mostly for use on large municipal buildings. It is good to know that a product of Lakeland is so eagerly sought after. The Canadian Imperial Bank of Commerce in Montreal used 125,000 square feet of naturally riven Spout Crag silver grey/green to add beauty to the exterior of this magnificent building. The St Paul's Tower in Dallas, the Connaught Center in Hong Kong, the City Centre in Mainz, an office block in Denver, the library in Copenhagen, the Old Bailey in London, St Catherine's College in Oxford, and the Holiday Inn in Glasgow are just a few other major buildings to use this fine slate. It is so durable and will last for ever. Much of it is polished and used for memorial stones and as surrounds for fireplaces, clocks, barometers, and many fancy items.

At the turn of the century Langdale quarries employed about 150 men. The quarries in Great Langdale were Thrang Crag, Low and Lord's quarries, and Spout Crag which yields a hard slate which polishes beautifully. There was also Banks, Colt Howe (known as 'Couta'), and Lingmoor quarry which has not been worked for years. The 'hundred steps' up to it are interesting but are now in a bad state of repair. This quarry was so high up that the slate would have to be carried down by hand, possibly by horse and sled, but it might have been too difficult for the horses to negotiate.

The local quarrymen had their own names for the different quality and sizes of slate, grading them into about 20 groups. In terms of quality the main ones were Best, Second Best, First Peg, Second Peg, and Tom.

The Little Langdale quarries were mainly on the Tilberthwaite and Coniston fells and are very numerous. Not many of them are worked at the present time, but years ago they were a hive of industry. Some people think that these quarries should not be worked as they are an intrusion into the valley. I do not see it like that. They create work for the local lads, and the stones and slates are very much sought after and are long-lasting. However, I do think more could be done to cover the old spoil heaps, although to me, they are no more than a natural scree.

Slate Worker in Trouble

Betty Yewdale lived with her husband in Little Langdale, but on this occasion, he had remained away all night after a funeral in Coniston. Betty set off to find him, and this is her friend's account of her return, originally told to her doctor, Craig Gibson:

> On i' t' efternen sho co' back drivin Jonathan afoar her wi' a lang hazel stick, an' he sartly was a sairy object. His Sunda' cleeas leuk't as if he'd been sleepin' i' them ont' top of a dirty fluer. Tie of his neckcloth had wurk't round till belat ya lug an'l lang ends out hung o'er ahint his shoulder. His hat had gitten bulged in at' side, an t' flipe on't was cock't up beeath back and frunt o' togidder, it wod ha' been a querely woman body at wod ha' teean a fancy till Jonathan that day. Says I, 'Ye hev fund him then.' 'Fund him! says she! 'ey, I fund 'im. I knot whar ut lait 'im. I fund him at t' Black Bull wi' yower meeaster and a lot meear o' t' seeam soort. They war just gan to git ther girt pan o' beef steeaks set on t' middle o' t'eeable. I meead t' frying pan an' t' beef steaks flee gay merrily oot o' t' door, an' I set on an git them sic a blackin as they will'nt seeune forgit. Then I haillt Jonathan oot fra among them, but when I had gitten him out wi' me I shamt ut be seen on t' roads wi' him. Sike a pictur ye niver see the like of. We hed to teeak t' fields for 't an'as it warn safe to let him climm the t' walls, I meead him creep t'hog hooals, an' when I gat him in, his heead in an' his legs out, *I dud switch him* (Told in Bradley, A.G., 1901).

Spinning, weaving, and felting

The crafts of spinning and weaving had long ago been carried out in quite a big way. Local dyes were used: certain lichens produced red or scarlet, oak apple gave black, broom – yellow, and birch bark – yellow or brown. The resulting cloth was carried to the fulling mills, formerly over the fells to Grasmere before the two fulling mills were built in Langdale at Millbeck and possibly at Skelwith Bridge. Where fulling was to take place there was always bracken and areas were cut in rotation by the locals year by year while it was still green. It was then burnt in 'potash pits', and the ash, mixed with lye and lime, was added to animal fat to make a soft soap which they sold to the fullers.

After dyeing the cloth was fulled to shrink it and to make it more durable. To do this it was put into wooden troughs and immersed in the soapy solution and pounded by wooden hammers driven by the water wheel. After fulling it was washed in the nearby river as in Mickleden or in the Tenters' Pond at Loughrigg and then put out on tenters to stretch and dry.

Felting goes back thousands of years. The wool is first carded and the streaks of wool are laid first one way, and then the next layer across the first, gradually building up the thickness to the required amount, different coloured strands making attractive effects. When the felt has been built up it is then put into hot soapy water and then rubbed and pounded vigorously using your finger ends or a piece of wood, or it can be put between rollers. It will then shrink and become solid. I have made lots of things from felt – hats, waistcoats, tea-cosies, egg-cosies, and children's mitts.

Spinning and weaving had almost died out until, in 1876, Albert Fleming of Neaum Crag started to revive the crafts. He found a few old spinning wheels and one or two elderly spinsters who were skilled in the craft and got them to teach other women. The base was at St Martin's in Elterwater, and was run at first by Miss Twelves. She had an exhibition in Doncaster in 1885 when she created quite a sensation with her wares and the demonstrations of spinners at work. It was later taken over by Elizabeth Pepper. The product was called 'Langdale Linen'. She wrote this poem in 1890:

Langdale Linen

In a little village 'neath the pikes,
There's some industrious people;
Aye! Some that live beneath the shade,
Of Brigg How's old church steeple.

In passing quietly through the vale,
You hear them gaily singin'.
You set aside to see them spin
Fine thread for Langdale linen.

In days gone by, when t'wheels were mute,
And they were getting rusty,
There were just two who thought about
Reviving this industry.

Twas work of labour and of love,
To set the wheels a spinnin';
But now you'd be surprised to see
The progress of the linen.

One fine old dame peeped o'er her specs;
'Whoa mun we thank for t'spinnin?
What wod it be if it wasn't for
John Ruskin and Albert Fleming.

We are grateful for these gentlemen,
For helping out our living,
By helping those who'll help themselves -
'Tis better far than giving'.

Now, if you want the full address,
And where to buy the linen,
St. Martin's, Langdale, Ambleside,
There, too, you'll see the spinnin'."

Lizzie Pepper Feb 18th. 1890.

In recent years a spinning class has started at Skelwith Bridge of which I am a member. It is a flourishing group with sometimes as many as 16 in attendance. As well as spinning, a variety of crafts are learnt including weaving, inkle-loom weaving (for making braid for braces and the girdling for horses' harness), tapestry, and the ancient craft of felting. We learn worsted spinning as well as woollen; also how to sort and blend the wool. This, too, is a very interesting craft. On the one hand you can leave the wool in its natural colours which is very effective, but we also learn the art of dyeing which is mostly done with plant dyes. All sorts of plants are used such as bracken, dyer's greenweed, nettles and the like. You also have to use a mordant to fix the dye. No wonder the old folks enjoyed it and in the olden days this had to be done, as well as knitting all sorts of garments, socks, and stockings.

Stained glass

Bernard Dean Walmsley and his wife Beatrice came from Lytham St Annes on cycling holidays to paint the local landscapes, staying at Mawson's shop in Chapel Stile. They soon fell in love with the area and after a few years decided they would like to build. Their favourite position was the site of the present St Annes which was then a field. Tommy Greenhow of Harry Place owned the land but rarely spoke to them. However, in 1929, having learnt that Beatrice came from country stock, his attitude softened and he unexpectedly said 'I 'ear you want this land'. 'Yes' said Dean, 'I want to build a house here'. 'You can have it for 40 yowes (ewes)' retorted the farmer. In calculating the price of the plot he reckoned that area of ground could support 40 sheep in a year. Tommy was unwilling to enter into any written agreement, claiming that his word was his bond. Several encounters subsequently took place between Dean and Tommy before he eventually came round to sign a deed of sale. Dean added the proviso that he should have the first refusal on the adjacent land, a very much smaller piece, which the farmer later sold for 30 ewes – thus upping the price!

In 1931 he built himself a magnificent house which he called the 'St Annes Studio'; the architect was Jonathan Stables, my uncle. It has a commanding view of the Langdale Pikes and a beautifully laid-out garden.

Mr Walmsley was an expert in making stained glass, and formed an excellent partnership with his wife who was a great artist, executing all the original designs for the art work for the windows. The window in Great Langdale church was created by them in 1937-38 and depicts St Francis of Assisi. It is remarkable for the light and colour entering through it on a sunny day and for the local wildlife – deer, red squirrels, goldfinches, bullfinches, swallows, and so many other lovely details including the Walmsleys' dog, Jimmy.

Their work is to be found in Salmesbury Hall and many houses throughout the country and in the Isle of Wight. Another magnificent example is to be found in the Elizabeth I Grammar School, Blackburn, in a series of ten Elizabethan windows depicting Drake and Raleigh and other characters of those days.

Beatrice, who was very gifted in embroidery and in making jewelry, died some years ago but Dean lived on until 1985 when he died at the ripe age of 91.

Sticks and crooks

Very rarely does a shepherd go into the fells gathering sheep without his crook. It is a help in walking, like part of your essential clothing. You can use it to hook round a branch to pull yourself up, or if your joints are getting stiff it is a great help coming down the fell. It is the ideal height for leaning on, while you rest or talk to a friend or for steadying yourself in the wind as you watch your sheep. It is vital for catching sheep – a shepherd is no good at lambing time without his crook – mind you there is an art in catching a sheep with a crook. You never see a shepherd at the sheepdog trials without his stick – either leaning on it or using it like an extended arm, guiding the sheep into the pen. I remember hearing two shepherds talking and one was admiring the other's stick. His little boy was there and the shepherd asked him about his stick – the little boy replied 'I don't wear a walking stick'!

It is a wonderful piece of equipment which has evolved to a basic pattern and shape, and is useful in so many ways. It is not surprising that over the years on long winter evenings people have whiled away hours making their own crooks and sticks and that these have become one of the features at the different shows. I have enjoyed making many sticks and crooks in my time.

We use rams' horns, Herdwicks or 'Swaddles' in Langdale – a really good horn is hard to come by. First of all you cut it off where it is solid and then rasp it down until it is smooth. Then you put it in a pan and boil it up for an hour or so, in order to soften it so that you can twist it into the rough shape. To do this you put the heel into the vice and turn and twist the rest of the head until the curve and the alignment are approximately right. At this stage clamps and vices are needed to hold it into the required position until it has cooled down. If there are small defects, bends or bumps in the wrong places, you can heat that part with a Bunsen burner or gas burner or over a lamp – taking care not to burn it. The next stage is to select and prepare a suitable hazel stick, straightening it by heat and pressure, before boring a hole into the heel of the horn. It is good to insert a nail into the stick and into the heel before glueing the two parts together, taking great care not to damage the bark at the join. Now you can rasp off the parts of the horn you do not need and shape it just as you want. Every stick will be different and you have to be guided by the shape and thickness of the material you have to work with. It takes hours to rasp down and then, with glass or sandpaper, carry on to a fine finish and finally treat with wax polish or varnish.

Some people are very skilled at carving animals on the head – foxes, sheep and sheepdogs, trout and pheasant heads. I have judged sticks and crooks many times at the Rydal Sheepdog Trials and the Coniston Fayre. Each judge has his own preferences – I like to take hold of the stick itself to see if it is absolutely straight and well balanced, to note if the stick and head have been neatly joined without cracks and have been finished well, to check the width of the mouth which should be wide enough to take a man's hand and therefore a sheep's head or neck, the nose of the stick should be aligned perfectly without twisting off line and I, personally, prefer a square heel. The sticks are arranged in different categories – fancy or ordinary walking sticks, and shepherds' crooks, plain or decorated. The standard is so high and the entries so varied that it is very difficult to judge which is the best.

Shoemakers

In 1692 Anthony Harrison, cordwainer (shoemaker), was in business at Harrison Place, now Robin Ghyll. Years later, around 1830, Henry Stables (my great grandfather) had a thriving shoemaker's business at Walthwaite; he lived to the age of 96 and died in 1906. This business was carried on by his son John. Two of John's sons were also shoemakers. One, John, died at the age of 34 but had already started the shoemaker's shop in Ambleside which still bears his name. The other son, Harry, was also a very good shoemaker. He built what used to be the Post Office in Chapel Stile. He died in 1925.

Harry Stables eventually stopped shoemaking and built himself a garage and kept one or two cars as taxis. He also installed the first petrol pump in the valley which used to be outside the old Post Office. The garage was made of wood and eventually belonged to me until I sold it to George Long who pulled it down and built himself a house on the site.

Joe Gaskell had a shoemaker's shop at Elterwater situated at the end of the Britannia Inn. He delivered telegrams up the valley on his bicycle. Jim Haughey, at the Co-op in Chapel Stile, was another shoemaker. He was quite a character with a real sense of humour. He never swore but used to say regularly ‘by the flames’. A visitor took a pair of shoes to Jim for repair, but they were really irreparable. Later, when she went to pick them up, she asked ‘Are my shoes done?’ He replied ‘They were done when they came’!

Old Jim was quite a character. He liked a few pints of beer. He would go to the Old Dungeon Ghyll Hotel for a pint and then trot down to the New Hotel – he always seemed to be on the trot. Then he would wend his way down to the White Lion in Chapel Stile for another one. He was really taking his little terrier, Peggy, for a walk. When he was in the New Hotel he invariably landed in the kitchen and tormented the maid servants but if Mrs Fothergill saw him she would just stare at him which was enough to do the trick. Once when this had happened he said ‘She’s a nice woman, is Mrs Fothergill, but I don’t like her’!

Fred Bowness was the shoemaker in Little Langdale while John Rigg was the clogger in Chapel Stile. Clogging was a different trade being less skilled: a clogger would not normally be able to make shoes and boots.

Tailors

In the year 1754 Jacob Fletcher, a tailor, had a business at Oak Howe. Latterly, this trade was carried on by Mr Tyson and by Benson Taylor who lived at Undergarth. He measured me for my first ‘fustins’ – fustian breeches (cord trousers) – after I left school in 1924. He would make all sorts of clothes, jackets, trousers, and suits. ‘Bense’ was a great character and always removed his false teeth to eat!

Shops and Post Offices

From time to time there have been numerous grocers’ shops in the valleys. The Great Langdale grocer’s, in the lane just above the White Lion, was run by Mr Jackson and Mr Grisedale and, when I was young, by Jim Thompson, whom we called ‘Old Shindy’. He had a marvellous bass voice and sang in Langdale church choir for several years. All his family of seven boys and two girls were musical. His daughter Lizzy married my Uncle Harry. Fleming Mawson took over the shop and was a man of great knowledge and a leading member of the Langdale community. He was a churchwarden and with his wife ‘Aunt Jane’ worked very hard for the church and parish. One project was to provide iron gates for the cemetery. John Hodgson continued the shop but it was closed when he retired and has been made into a private house, Orchard Cottage, where Dot Parker lived.

The Co-operative stood originally opposite the Glebe in a building owned by my great grandfather. When the present store was built just over a hundred years ago the original large building stood empty for a while and then was made into four cottages. The architect was Jonathan Stables, grandson of the owner, and the cost for the whole conversion in 1901 was £126.4s.4d.

Nelly Webb and her sister Violet put up a little wooden hut at Thrang Road corner where they sold all sorts of toffees, cigarettes, and soft drinks. Violet married and left the business; then when Nelly married Percy Youdall they built an extension for living quarters. The small shop was eventually closed and the whole building is now a cottage.

Miss Edith Earl started a small baker's establishment at No.2 Lingmoor View. As children my mother used to send us for bread when she ran out. She made excellent brown bread and I have never tasted any like it since she gave up the business! A Mrs Pickersgill ran a confectionery shop at Chapel Stile for a short while.

The Great Langdale Post Office was at Elterwater at the present Corner Shop and was run by Mary and Lizzie Thackeray together with a general store. Between times they were always busy doing embroidery which they sold. After they retired the Holmes family took the Post Office, moving it to the other end of Mill Row. It then moved to Chapel Stile for a few years but in 1992 returned to Elterwater.

Around the turn of the century there was a chemist's shop at Daw Bank. There was no butcher's shop as most of the farmers killed their own animals and some went round with a horse and cart selling meat. There was Tom Chew, the butcher in Grasmere, and George Cragg in Ambleside, both used to go round their areas with horse-drawn carts. George later sold the business to Asplens in Ambleside. He owned Harry Place – his wife was my grandmother's sister.

From time to time there have been a number of grocers' shops in Little Langdale; one was run by a Mr Greenup. The valley also has its own sub-Post Office.

Blacksmiths

There would be one or two blacksmiths' shops in the valleys. In Great Langdale there was one at Ellers. An old deed of conveyance dated 1693 registers the transfer of a small field with out-buildings near Ellers from Anthony Harrison of Harrison Place to Thomas Fearon of Robinson Place, part of 'George's Tenement' and called Shop Field. Here was a smithy as well as a joiner's shop. When the old buildings were removed in 1909 distinct remains of the smithy hearth and chimney were found next to the joiner's shop. My grandfather used to tell me that he used to 'bark' oak wood there – remove the bark off the wood to use it in tanning leather.

The blacksmith's shop in Elterwater I remember well, situated where Mrs Riley has her spinning shop. It was run by John Bowness who also ran the White Lion in Chapel Stile; he had a large family of seven girls, one of these, Janet, married our local vicar, William Hayes. The last blacksmith was Jack Clark. I have taken the horse to be shod many times. When the cart needed any work on the wheels it was taken to the joiner first. In Elterwater in those days it was William Holmes, one of Johnny Holme's fore-elders. After a number of years some of the spokes, coming from the 'naff' (hub), would need to be replaced, but the main trouble usually was that the 'fellies', forming the wooden rim, gradually rotted and then the hoop became loose. William Holmes would make and fit the new spokes and fellies and then take it to the blacksmith to be hooped. The Smiddy Dub by the roadside in Elterwater was dammed to collect a sheet of water and then the hoop was heated on a fire by the dub. It had to be fitted onto the wheel while it was red hot, and then plunged into the cold water of the dub. This contracted the iron hoop to fit tightly onto the fellies. This smithy closed down in the 1920s and then we had to take our horse to Grasmere to be shod.

The blacksmith was a very skilled and important person, who could not only shoe horses and make and repair cart wheels and farm machinery, but could also make tools, iron work, millstone bands, nails, hinges, leys (scythes), axe heads and billhooks, to name a few. It was a sad day when our own Elterwater blacksmith's shop was closed down.

The Smiddy Dub gradually silted up and rushes have grown up in it – it looks unsightly in its derelict condition. However, in 1989 the Langdales Society contemplated dredging the dub and restoring it to its former state. The Gunpowder Works had their own smithy, away from the works to avoid any sparks, up on the hillside opposite the main entrance to the present timeshare complex.

Joiners

Apart from the joiner's at Ellers, the joiner's and undertaker's business at Elterwater was run by the Holmes family, firstly by William, then Joseph, then Jackie, and finally by Johnny, who still has all the old work ledgers and the records of funerals. The funeral records include the diphtheria epidemic of 1913, the first burial in the new cemetery, and the effect of the explosion in the Gunpowder Works in 1916. This business was closed down a few years ago and the premises made into cottages and flats. The joinery firm of Brown & Long now has a flourishing business and employs a number of men, they also carry out building work and are the only undertakers in the valley.

The specialised work carried out by Joseph Holmes included making all the looms for Albert Fleming, making the water wheels and restoring the sluice at the Gunpowder Works. They made all the quarry ladders for Great and Little Langdale, Tilberthwaite, Coniston, and Honister quarries, also the 'gants', small, square, moveable platforms which were attached to the ladders and on which two men could just stand to bore the holes for blasting, often about 40 feet from the ground. They made the slate sledges for working at the top of Lingmoor and Honister where no horse-drawn transport could reach. One was dragged on foot by Bill Hartley and one or two workers who lived in Chapel Stile up Mickleden, Rossett Ghyll, and on to the Honister quarries! They made flat-bottomed boats for Colonel Robinson and for Mr Balme. Being the wheelwrights, they made carts and wheels for the farms and complete traps and waggonettes.

Cycle shop and garage

About 80-90 years ago Robert Smith started a cycle shop in Chapel Stile before moving to Windermere where he created a garage business which has gone from strength to strength. Bob Smith was the same age as my father and they went to school together, they were both good scholars and my father only beat him by a few marks to win the place at St Bees School.

After Cyril Clark came out of the Army in the early 1920s he bought a chara motor (charabanc) with doors in the side which he called Lady Isobel. He had it garaged at the coal houses which have now been made into a house and gift shop now in the occupation of Mrs Huddleston who sells many interesting items of local craft. Cyril Clark had the first gramophone in the valley – with a large horn, one of the original type – and he brought it many times to Pye Howe for us to hear. Later we got one for ourselves, but without the horn!

Callers – travelling salesmen, pedlars, and tramps

We often had pedlars and hawkers calling on us selling their wares, carried in a basket or box fixed on each side by pieces of string or a strap and hung around their neck. Some brought cotton, braid, tapes, pins and needles, others offered moth balls while others sharpened scissors and knives. One I remember well was Sammy Morgan, he often had his wife with him, and there was a Scotsman dressed in a kilt – he used to come and sing!

The occasional hawker would come with a horse and flat cart and the potters and tinkers would call buying rags and old dirty broken wool – you might get about 2d. for a bag of rags and about 2s. for a bag of broken wool. The hawker, usually a gipsy, would exercise his whippet dog and let the horse eat by the roadside.

A Mr Huggon used to travel with patterns of materials from a firm in Kendal called Graham's. He carried his patterns in two packs all neatly packed in two black pieces of oilcloth; each was about 2 feet by 1 foot, fastened with a strap and carried over his shoulder. He always walked round here. No doubt he would come by train to Windermere. He would come in and spread his patterns out on the table and mother would order what she required. He carried all sorts of patterns – tweeds, linens, cottons, and flannels. The materials you ordered would come later on and we used to pick the parcel up in Chapel Stile. Then mother would go to work making clothes for us. Parents made most of the children's clothes and their own, including the menfolk.

We also had the men who used to come round with straps of onions – often they were Frenchmen and with a bit of bartering sometimes you got the price down.

In the olden days it was quite common to see a tramp or vagrant. These people would travel from house to house begging, sometimes for money or a bite of bread and cheese. They often carried an old tin with two holes pierced near the top for a wire or string to hold it by and you would be asked to fill it with tea. They often begged for a clean garment – a shirt or socks – often they were wet through and asked if they could sleep in the barn. We would let them in on the surrender of their pipe and matches – this was for fear of them setting fire to the barn. As children we were a bit afraid of the tramps, but I don't think we need have been as they were harmless enough. They mostly seemed to enjoy the life and didn't seem to have a care in the world. Some of them were quite learned people and no doubt had seen a great deal of the countryside of England.

All these customs seem to have died away, you hardly ever see a tramp nowadays, nor even a hawker. Sad, because this was part and parcel of country life.

Robinson Place

CHAPTER 7

TRANSPORT AND COMMUNICATIONS

Shanks's pony

Going back many years horses were the only means of transport – if a horse was not available you had to travel on foot. My mother once told me of a woman from Langdale who walked to Kendal to do her shopping, a distance of 25 miles each way, a marvellous achievement – presumably she stayed with friends overnight. We always had to walk to school and some of the children would walk over three miles each way. The menfolk also had to walk to work and in some cases it would be five or six miles and they had to be on the job at 7.30am. It did not seem to bother any of us as it was all taken in our stride!

Our neighbour, John Rigg of Chapel Stile, on Monday, 10 May 1880, started to keep a record of his daily trampings to Newton Quarry and back. On the same day in 1901 he completed his 21 years employment at Messrs Newton's and as a matter of curiosity added up the number of miles he had walked in pursuit of his daily bread. He found he walked no less than 6,395 times or 23½ times on average per month. As the distance is three miles each way, it will be seen that he travelled no less than 38,390 miles – which is equal to walking one and a half times round the world, or to walk to London once every two months during the 21 years. Well done, John!

Motor transport has spoiled all of us, yet even now lots of men are late for work. I think the reason is that most people stay up late, not going to bed much before midnight and then they cannot get up in the morning! Years ago everybody went to bed early as there was not much recreation. In summer, when the days were long, and at some times of the year the farmer would work as long as the daylight lasted.

Horse-drawn vehicles

Horses were of great value before motorised transport came. The packhorse was often used to carry big loads over long distances, and also for riding. Horse-drawn sledges were used for carting bracken, peat, and stones for walling, as was the horse and cart. About 60-70 years ago the hotels had four or five horses and had two-, three-, or four-horse 'charas'.

Some also had waggonettes, phaetons, landaus, and broughams. The phaeton was a four-wheeled open carriage drawn by a pair of horses with two seats facing forward; the waggonette was four-wheeled, open, with a removable cover and rows of benches along the side facing inwards; the landau, originally made in Germany, had four wheels with a top in two parts which could be opened; the brougham was a one-horse carriage with two or four wheels for two or four persons.

In our young days the horse charas had a regular route up Little Langdale, over by Blea Tarn and down Great Langdale. They had to negotiate gates across the roads at various places to stop the sheep trespassing off the common. There was a gate by Hacket Forge in Little Langdale, one on the road to Wrynose just after the turn off to Blea Tarn, and one by Blea Tarn. The one I remember most was at Harry Place and was in use as late as the 1920s. Near the gate there was an old water trough which is still there. It is a square trough hewn out of a single piece of freestone and supplied by a spring of lovely cool water coming out of the bank nearby. When we were coming home from school we often stopped to have a drink there. Along the roads there were lots of these troughs at intervals down the valley. Most were very primitive and often made near a spring on the roadside out of a few flat stone flags. These were most important as you must not let a horse drink too much cold water at once as it causes colic. I think these old troughs should be preserved as they are a lovely feature of the valley. There was another gate at the top of Langdale Common but this was missing for several years and was not replaced by a cattle grid until about 12 years ago. Before this sheep and cattle wandered down the road to Skelwith and were quite a menace to cars and waggons, and many of the animals were killed. Earlier all these gates were manned by young children who, no doubt, made a tidy sum of money.

The long vehicles with seats facing forward were light enough to be hauled over Blea Tarn Pass. They would stop for tea at one of the hotels, calling in at 3.30 or 4pm – it was like a bus service and you could make a trip to Keswick on them. Mary Birkett, who used to live at Wall End, well remembers seeing the charas at the top of the pass and hearing the coachman sounding his horn to inform the staff at the Old Dungeon Ghyll Hotel to put the kettle on and prepare for the visitors!

The farmers all had dog-carts. When my father was going to Kendal he would go in the horse and trap to Windermere Station, leave them there, and go on by train. It was a day's work. If the roads were slippery you had first to take the horse to the smithy to be 'sharped', that is to have special spikes added to its shoes. It was hazardous when the roads were bad in winter – sometimes you would have to creep down the side of the road using the verge. You would, of course, only set out if it was absolutely necessary.

If the cottagers wanted some coal the farmer had to go to Coniston or Windermere to collect it for them. When all the farmers went to the Keswick Tup Fair to buy or hire rams, they often hired a horse chara from the hotel, but they would have to walk the steep hills like Red Bank. It is all so simple now as most farmers have Landrovers or vans.

The horses were very good and patient and would usually wait wherever you left them, this was very useful when you were haymaking or delivering goods. One local farmer, who liked his drink rather too well, one night parked his horse and trap outside one of the inns in Langdale. He got a little worse for wear so one of his friends slipped outside, unyoked the horse, passed the shafts through the gate and harnessed them up again. On leaving the inn, the farmer was extremely puzzled and ran back inside exclaiming 'Come quick – horse 'as lopped ge-at'!

Horse transport was one of the most important means of getting about or moving anything – coal to furniture or anything else. I remember my mother telling me that she left school when she was 12 years old and that, on the very next day, her father sent her to Coniston Station for a cart-load of coal for a neighbour. She cried because she didn't know the way, but her father made her go saying 'The horse will show you the way' and it certainly did! After that she went numerous times – the cottagers had no other means of getting their coal. One day Thomas Youdell of Low Millbeck went to Coniston for a cart of coals. You could get coal either at the station or from Eldy Mossop. Thomas always got them from the station master but on this day he didn't have any left. He went on to Eldy and asked if he had any, but Eldy replied 'I have plenty of coal but none for you!' Thomas had to come all the way back without any.

During my young days a baker used to come round with a horse-drawn van selling bread and cakes, and a butcher used to come in a similar vehicle. When the Allonbys were at Baysbrown, Dora used to go round the village with a milk-float and milk was measured out into a jug. Other farmers carried milk round in two big churns, also with a measure. Now most of the milk is collected and goes to the big dairies to be pasteurised, coming back in bottles or cartons. To me it is never the same as milk straight from the cow.

When I was five years old my mother took me for a trip round Thirlmere in the horse and trap. We took food with us and had a picnic en route. This was our means of pleasure in those days and we greatly enjoyed ourselves. We must have taken someone else with us as I can remember that the trap was full and my sister Margaret accompanied us on her bike. Maybe this was the first time I had been far from the doorstep.

After the penny-farthing bicycle, which must have been quite tricky to mount, came the more useful bicyle with two wheels of the same size. This means of conveyance was extremely good and we used it a lot when driving sheep or cattle. Even if you couldn't ride when you were driving them it was very useful for returning home. At night we had to have a light; the first ones were little paraffin lamps but these were easily blown out. The best were the carbide lamps, but if you were going any distance you had to carry some extra carbide in your pocket. This was before electric torches came along. Later on I had a dynamo which was very efficient.

The postman came in a horse-drawn vehicle and sometimes as children we got a lift. When he got to Wall End, the end of his round, he would stay and help with the farm chores. He was given his food as remuneration for the work he did. Then he went back to Ambleside and collected from the boxes all the way down the valley. In later years he came on his bicycle with one large holder on the front and one on the back; after this he had a motor cycle with a side-car box. Now he comes in a van. We used to get our mail much sooner up the valley in those days. I suppose there is so much more mail now, but I don't know whether this is progress.

Telegrams used to come to Miss Thackeray at the Post Office at Elterwater. She would go across to old Joe Gaskell, who had a shoe-shop at the end of the Britannia Inn, he would get out his push-bike and deliver the message. If he had to go to the top of the valley it was quite a slow means of communication.

Telephones

The telephone came up the valley in 1923 and it was really something. At first the wires were on poles all the way up the valley bottom; later they were put underground. If there was heavy snow it brought down the wires – there were wires everywhere. There was chaos.

Parish outing – 1894

At the turn of the century if you were going on a choir trip or some such excursion the horse-drawn vehicles would convey you to one of the railway stations, either Coniston or Windermere, so that you could catch the train. The Langdale parish magazine for October 1894 carried an account of one such excursion:

> The Annual Excursion of the Langdale Church Choir and Sunday School Teachers came off on Friday, September 7th, the destination being Morecambe. The rendevous was Dog Kennel Corner at 6.15am and by the appointed time the whole of the party to the number of about forty had turned up, some of the seniors looking, it must be confessed, a bit sleepy, but the high spirits of most of the party quickly chased away the morpheus-like feelings. The party was soon en route to Windermere in conveyances supplied by Messrs Baines, Bowness and Youdall in time for the 8.10am train from there. The station was reached in good time, and all were comfortably seated in a carriage kindly provided by the Company for the exclusive use of the dalesmen and the journey

to Morecambe was soon an event of the past. The headquarters at Morecambe were the Winter Gardens where an ample spread was keenly enjoyed at mid-day, the good things having been largely given by kind friends at home. The day was spent all too soon, donkey rides, tricyle rides etc. filling the gaps between the various entertainments at the Gardens which were enjoyed to the full by the entire party. At 5.30pm the party again assembled for tea and, at the conclusion, all were asked to meet at 7pm for Roll Call before proceeding to the station. On arrival at the station the packing and numbering were carefully performed by the Vicar and the return journey commenced. The way was enlivened by various songs and catches etc. and, although one or two were disposed to sleep, the hilarity of the rest effectually prevented this. The carriages were waiting at Windermere and very quickly indeed the whole party was safely wheeled home having thoroughly enjoyed this, their eleventh trip.

Octogenarian on Scafell – 1891

My great grandfather made headlines in the local paper when he climbed Scafell in his 82nd year in 1891. It was reported in this way:

Ascent of Scawfell by an Octogenarian

With the best of weather, health and strength, the ascent of Scawfell is an undertaking at any time but, when a man is handicapped by upwards of the allotted span of years, the feat becomes one well worthy of record. This memorable feat was, however, accomplished on Monday, 15th. inst., by Mr Henry Stables, Chapel Stile, Langdale, now in his 82nd year. He was accompanied on the occasion by his son, Mr. John Stables, Walthwaite, his grand-daughter, Miss Hannah Stables, Walthwaite, and Mr. Whalley, manager for the Langdale Co-operative Society, so that another noteworthy feature is that three generations of the family were represented. The distance from Chapel Stile to the summit was traversed in under six hours, the old man tripping along with the best, up and over the difficulties of Rossett Ghyll, where many a good intention has been frustrated before today, leading the way instead of being led, and only occasionally pausing to enquire how his companions were progressing because, he afterwards said, he was afraid he would have some of them to carry before he got home again.

The summit was safely reached before noon, and rest and refreshment indulged in, while the peerless views of mountain, lake and valley were enjoyed for an hour or two, and then a start was made for home. On the descent the veteran was much amused at meeting a party of tourists in patent leather and top hats – rather a strange outfit for Scawfell and one which would be likely to bear the marks of pedestrian warfare for many a day. Chapel Stile was reached in the early evening and the old man turned up as fresh as a lark and with some notion of acting as a guide in the future for, he says, 'It's nobbut a canny ebenin's walk'.

The following year he went to see a friend in Wasdale and, returning via Eskdale, he was so little tired that he went into the hayfield and assisted the family for more than an hour. Even later he walked home from Keswick, Cartmel, Windermere and other places.

Another article appeared in a local newspaper in 1896 when he was 86, nearly 87, and for sheer toughness and tenacity of spirit surely he takes some beating:

Henry Stables is a notable example of vigorous old age. He was born on December 1st

1809 in Great Langdale. He married Agnes Taylor by whom he had ten children, five boys and five girls. For many years he followed the occupation of village cobbler, he also farmed some land. At 75 years of age, according to his parish priest, he would jump an ordinary five barred gate by simply placing one hand on it rather than take the trouble to open it. His most wonderful power of vitality was shown on Jubilee Day 1887 when he was 78 years old. Rising at 6am he started for the Langdale Pikes nearly three miles distance. He assisted in carrying up material, about 18 lbs in weight, for the bonfire at the top of the 2,400 ft. mountain, reaching the top at 7.30am. After resting, he returned home and joined the 'old folks' at dinner at 12.30. He took part in the Jubilee Service at 1.30pm, thence went into the sports field and won a race of 200 yards against many competitors of 50 years and upwards and stayed on the field until 9pm.

Not content with this, at 10.15pm it seems he set off for Loughrigg to see the Jubilee bonfires, stayed there until midnight and reached home at 1.20am. Next morning he was up at 6.30am to drive his son and grandson to Coniston en route for the Isle of Man.

In commemoration of these feats he was presented with a Silver Jubilee Medal by the Honourable William Lowther, MP.

Pye Howe

CHAPTER 8

SPORT AND LEISURE

Langdale Gala

Most of the valleys have their own major social gathering – Wasdale has a shepherds' meet where the different farmers and shepherds gather to sort out lost and stray sheep. To these occasions are added all sorts of competitions and sometimes hound trailing and sheepdog trials. Langdale used to have a gala, called the Mechanic Club amd Langdale Club Walk and Gala, held on Whit Friday. People would assemble in Elterwater and walk to Chapel Stile, led by the Langdale brass band, for a service in the church followed by dinner in the White Lion room, across the road from the old White Lion. Everyone would make their way for the gala in Baysbrown field – it is still called the Gala Field to this day – for the sports which included fell-racing and wrestling. At night there was a dance in the Co-op room or in the Drill Hall in Elterwater before it was sold and made into a cottage. With the outbreak of the Second World War the gala ceased to function. I clearly remember just after the First War, in 1919, there was a children's sports held in Elterwater Meadow followed by tea in the Drill Hall

The brass band was formed in the autumn of 1892. In six weeks the instruments had been bought, the rules drawn up, and the first practices of the Langdale Subscription Band had taken place. They were all ready to head the procession at the next gala.

The Langdale Gala was revived in Jubilee Year, 1978, to be a family and community fun day and continues to the present day. A Gala Queen is chosen and a fancy dress parade is held. Track events and fell races are organised for children and others, judging of sheep and examples of local crafts, home produce, and plants are other attractions. Each year there are demonstrations of local skills – sheep-shearing, shoeing horses, and riving and trimming slates. All this is to the background music from a Gavioli giving a fairground atmosphere. Later each year Stool End Farm holds an open day with many similar local attractions.

Fell racing

We in Langdale produced many first-class runners who competed locally and in the Grasmere Sports. John Greenop, who was born at Walthwaite, was a famous runner, winning the Grasmere Guides' Race six times in succession, from 1876-81; he was third in 1884 and eighth in 1889. He was born in 1854 and died at the age of 70 years. His brother William was second in 1869 and first in 1870 when he also won the one-mile race, and was third in the Guides' Race in 1875. Another remarkable runner was George Woolcock of Weir Cottages, Chapel Stile. He won the Guides' Race at Grasmere in 1920 when he was only 17 years old, and repeated his success in 1921 and 1922. In 1920 the *Westmorland Gazette* reported:

> Langdale this year has produced an exceptionally good fell runner in George Woolcock who has now secured ten wins to his credit in fell races this season, including Grasmere, the premier event in Lakeland, thus establishing a local record. Woolcock, who is only 17 years of age, is a well built lad of medium height, and possesses an extra turn of speed, great stamina, and is sure footed – qualities so essential to success in fell running. This is Woolcock's first season on the fell.

He was only 5 feet 7 inches in height and weighed 9 stone. His knowledge of fell craft was gained by following the local foxhounds.

My great uncle, Fletcher Harrison of Ghyll Foot, Grasmere, was a shepherd and a keen fell runner but his mother did not like him taking part. On the day of one Grasmere Sports she kept lunch back until it was almost too late for him to attend. He managed to get hold of a jug of butter milk, drank the lot, and went on to win the Fell Race!

Wrestling

Cumberland and Westmorland wrestling was a male sport enjoyed by both young and old. In my younger days the boys learned to wrestle in the playtime between school lessons. It has largely died out as a school sport but thankfully it has been revived and the lads can go to classes again – it is based on strength and skill, not on violence in any way. It is one of the most popular events at the Grasmere Sports. The men wear fancy costumes, beautifully embroidered by the womenfolk. In a way this is not really necessary, but most besport themselves in their magnificent attire and a competition is held for the most attractive costume. In the real wrestling contests the men skilfully used the different 'chips and holds' – back heel, buttock, hype and swinging hype, cross-buttock, outside strike, hank, inside and cross click, and hold.

This old sport was practised two or three hundred years ago, but like many sports these days it has been spoilt by 'barneying' (paying money to let another win), and by the giving of big money prizes. The prize years ago was a fine leather belt – I have one of these in my possession, but not won by me! I don't remember any outstanding wrestlers from Langdale, but we always enjoyed watching it.

Local clubs

Old parish magazines record that the Langdale Cricket Club was revived in 1891; it played on the field below the Guide Post Cottage, Elterwater. The Football Club had not scored many goals in 1892 but was making a special effort to improve. It had several venues including Walthwaite Bottom Meadow and later the Gala Field at Baysbrown. The Cottage Garden Society held its annual flower show in August 1891 and prizes were awarded for cottage gardens. The Institute, in 1890, was fulfilling a need during the long winter evenings providing a billiards competition and other recreational activities. The Band of Hope, a temperance movement, was flourishing in 1897 when 130 sat down to tea at Mrs Crossfield's Club Room (the White Lion room), games and races followed.

Hound trailing

This is a summer sport and there are many keen followers in this area. Two men lay the trail, which can be up to 10 miles long, dragging a piece of rag saturated with aniseed. The two men start in the middle of the course, one goes one way and the other in the opposite direction. It is all timed for them to return to the starting point just before the hounds are released. The hounds are the same breed as the foxhounds but they are very well trained to follow the scent. At big meetings there are usually about four or five trails for different classes – puppies, old dogs, and maidens.

Each year there is always a meeting at Fell Foot in Little Langdale and often at Stool End in Great Langdale. Bookies are well represented at these events and the gambling seems to be very keen and the excitement intense. When a trail is run there is always a time limit and if the dogs come in either too early or too late the trail is adjudged to be void.

Often the gate money is collected for a worthy cause and the events are not solely for sport but are effective money-raising efforts. Many visitors to Lakeland are most interested in going to watch these local events, especially on a fine day! Sometimes the trails are held in the evening, but on some Saturdays as many as five trails may be held.

Fox hunting

Fox hunting is an old sport still carried on in the two valleys. It is a winter activity and is mostly done to keep the population of foxes to a minimum. There is no better way of controlling the fox in these areas than by hunting with a pack of hounds. Shooting would not be satisfactory; if a fox was maimed and not killed outright it would be sure to suffer a lingering death. When a fox is killed by the hounds it is an instant death. Poison would not be any good as this would be a hazard to other animals and, in an area like the Lake District, it would be a very difficult task to control the numbers other than by hunting.

The fox does an awful lot of harm to the farming community, in some years killing a large number of lambs and you dare not leave your poultry unfastened at night. Even at mid-day they are on the prowl and are becoming much more cheeky. On one evening no fewer than six foxes were surveying the possibilities round Stool End Farm. The packs of hounds are usually kept at one of the farms during their visits to this area. The foxes are always hunted on foot as horse hunting would not be practicable in these mountainous areas.

Great Langdale is hunted by the Coniston Foxhounds and both of the valleys by the Eskdale and Ennerdale pack. The latter pack was started by Tommy Dobson in 1857 with just two hounds, Cruiser and Charmer. He built his pack up to twelve and a half couples. He died at the Tourist's Rest, now the Three Shires, in Little Langdale on 2 April 1910. His body was carried over Wrynose and Hardknott Passes on a horse-drawn vehicle to be buried at St Catherine's church in Eskdale on 5 April. He was 83 years old.

After a good fox hunt with a kill it was the custom for everyone to meet at the local for a few drinks and a sing-song. The traditional hunting songs were sung and a real jovial night was had by all. This is what we call the 'arvel', from an Norse word meaning a feast or party, especially after a funeral or a fox hunt.

It was always the custom, when someone died, to have the arvel in the home of the loved one – with a good meal provided. The tale is told of a woman whose husband was on his death bed and not expected to last more than a day or two. It was fitting that Betty would put her man away with a fine fare. Markets for meat were few and far off, so she, being a good thoughtful soul, sent for the joint of meat before his death took place. John stuck to his breath. He would not die to save a joint of meat – so it went bad and had to be destroyed. She might perhaps buy another joint – but no. 'A'll dew nowt o' t' mack', says she. 'He sud a' deed befoor when he could have had a decent burial! He'll get nee a dinner frae me – for a'lbt send him off wi' a bit of breed and cheese fust'. Faithful Betty, she knew how John would have backed her up in her efforts to save both meat and mense (respect), and would have been the first to approve of her actions!

At the arvel many traditional hunting songs were sung, one of my father's favourites which he regularly sang was 'Drink, Puppy, Drink':

Here's to the fox in his earth below the rocks,
And here's to the line that we follow,
And here's to the hound with his nose to the ground,
Tho' merrily we whoop and we halloa.
Then drink, puppy drink,
And let every puppy drink
That is old enough to lap and to swallow;

For he'll grow into a hound,
So we'll pass the bottle round
And merrily we'll whoop and we'll halloa.

Otter hunting

Otter hunting was also carried on but is now banned. The otter never did as much damage as the fox and now they are very rarely seen.

Hare coursing

The harriers and beagles hunted the hares in Langdale years ago. The Windermere Harriers used to hunt in Langdale but not in this area nowadays. Hares are not very plentiful now. The hunting was just a bit of sport and the hare usually got away. When the beagles used to come to Langdale it was fun to see them as they were such short-legged creatures needing to be lifted even over a low wall – I never thought it was possible for them to catch a hare!

Cock fighting

In the early days cock fighting was one of the chief sports. Although it was banned in 1835 it has still been carried on in the dales, especially in the remoter parts. The cocks had their combs and wattles dubbed and were clipped for fighting trim. They were fitted with 1¼ inch spurs and were bred to fight with speed and ferocity. Big sums of money were gambled. This sport took place in most towns and villages and at race meetings. On Shrove Tuesdays the principal grammar schools held 'mains' as the fights were called, and the sport was followed by rich and poor alike. Thousands of gamecocks are calculated to have been killed within a radius of 10 miles of Langdale. Even today secret mains are thought to be held in isolated venues. My father used to breed gamecocks, English Game Red and also Grey, when I was a lad. They were fierce creatures and had a great big natural spur, so if they did attack you the best plan was to get out of their way! My brother was stabbed in the leg rather badly on one occasion. As far as I can gather there was no main in Langdale, but cocks were bred and the participants in the fight would travel miles for the sport – Dalston was one of the chief areas for cock fighting. There were many cock-fighting songs; here is one I can remember being sung:

The Bonnie Grey

Come all ye cockers from far and nare,
And I'll tell you of a cock fight when and where;
At Highfield Moor I've heard them say,
Between the Black and the Bonnie, Bonnie Grey.

Willie Woof in the garden at Robinson Place.

Herdwicks on Wrynose Pass, the road still unmetalled.

The valley road at Raw Head before widening in 1925.

Birk Howe, Little Langdale, 1947.

It's into the pub to take a sup,
And a cock fight it was soon made up;
For £20 those cocks will play,
The charcoal Black and the Bonnie, Bonnie Grey.

The first to come, it was the Oldham lads
And they brought all the money they had;
The reason why I've heard them say,
'The black's too big for the Bonnie, Bonnie Grey'.

Well the cocks they at it and one, two three,
And the grey was tossed – the Oldham lads cried
'Hey you're lost'.
As collier lads we turned quite pale,
And we wished we'd fought for a barrel of ale!

The cocks they at it and a one, two three,
The charcoal Black got stuck in the eye.
They picked him up and he would not play,
And the fight it went to the Bonnie, Bonnie Grey.

With silver breast and silver wing
He's fit to fight in front of a king.
We cry 'Hurrah, Hip, Hip, Hurrah'.
The fight it went to the Bonnie, Bonnie Grey.

Fishing

Fishing was another great sport using rod and line and night lines. An old method, which has been illegal for over 100 years, was 'lath' fishing. The lath was a piece of wood about 2 or 3 feet long by 1 foot wide, weighted with lead at one side and with hooks fixed at intervals. It floated on end and could be steered out into the middle of the tarn with a string. I remember going with my father in about 1922 to lath Stickle Tarn. There were fewer visitors then and you were pretty safe on a dull day not to see anyone. However, father had barely put out the lath when someone appeared. He had to draw it in and hide it under some stones, but when he returned some days later he could not find it – and he never did.

Thomas Myers, my great-great-grandfather, was very skilled in lath fishing and had been chased off the water's edge many times. He used to make his own flies and maintained that the fish were there to be caught – no matter how! Unfortunately the law said 'No'. Tom was also a keen follower of the otter hounds, a cock fighter, and a great man with a gun; he lived at Elterwater to the good old age of over 90 years.

CHAPTER 9

SCHOOL IN LANGDALE

Early days, new schools

Education was very scanty in the early days and I can recollect that one or two of the older residents could not read or write. Before the school was built at Brig Howe, there were one or two houses where children could go to learn – records tell of a little old building near the church which was used for teaching, and also a house in Elterwater where some children went. My grandfather used to tell us how he went for lessons at the old house at Ellers where the teacher was Will Spedding.

In the early days the vicar or curate or the 'reader' would give lessons in the church, but this would only be for the children with more affluent parents, and they would all have to pay their 'school pence'. In 1824 the Elterwater Gunpowder Company built a new school at Brig Howe in Chapel Stile, then part of the common. In 1860 Lord Lonsdale, lord of the manor of Langdale, conveyed to the school trustees the 2½ acres of land on which the school stood 'as a site for a school for poor persons of and in the said chapelry and for no other purpose whatsoever'. This building was only small and was enlarged on two occasions, the first in 1890. A further addition was made by Mr Wheatley-Balme to house the infants' department, which had for a time been meeting at Lane Ends. Part of this land was, however, put to another purpose when it was given up in 1898 for building the new road from the former Post Office to Thrang Road corner. The main road from Skelwith to Elterwater Common had then been recently constructed by the Gunpowder Company for their waggon hauls to Windermere Station, and they had also built the schoolmaster's house in return for permission to enclose more of the common.

Mr Forster's Compulsory Education Act of 1872 found the Langdales deficient in school accommodation for all their children, and so the Lecture Room in Little Langdale was immediately brought into use as a school. In 1894 the school in Chapel Stile could cater for 150 children and the average attendance was 120. The school in Little Langdale could take 60 children and the average attendance was 46. The population in the two valleys in 1891 was 754 in the township, 770 in the ecclesiastical parish. The Revd Fletcher Fleming gave additional land on which Wheatley-Balme built a master's house, while further ground was given by Colonel Robinson and Jackson Youdell of Birk Howe. On part of this land the new Little Langdale school was built in 1904-5 to the east of the mistress's house, so that the Lecture Room then entirely reverted back to its duty as a chapel. The cost of £800 for the new school was given partly by Mr Balme senior, 'In memoriam E.B.W.B.', and partly by local subscription, while slate and stone were given by the quarry owners and cartage was carried out by the farmers. This school was closed down in 1964 and now all the children come over to Brig Howe. The Skelwith school closed in early 1970 and the children from there also now come to Langdale.

School-teachers – Great Langdale

Samuel Twist	1859-64
C.J. Hopping	1865
C.D. Clark	1866-67
H. Sutton	1867-74
C.J.B. Ellis	1874-78
George Bentley	1878-83
Thomas Fisher	1883-1914
J.R. Barrow	1914-31
J. Hakitt	1931-37
Miss H.R. Wolfe	1937-43
Miss H. Chapman	1944-51
Mrs Langhorn	1951-71
Miss B.J. Tracey	1971-85
Richard Sanderson	1986-89
Nigel Tansley	1989-

My school-days

My days at school were quite strenuous as Pye Howe, where we lived, was 1½ miles from Chapel Stile. We had to walk – hail, rain, snow – many times we were wet through and I can recollect Mrs Barrow taking me into her house to lend me some of Bob's clothes to put on while she put mine to dry. Mr J.R. Barrow was our headmaster and his wife taught as well, and Miss Gaskell taught the infants. The Barrows had two daughters and a son, Bob, who was older than me so his clothes were a bit on the large size, however, this did not matter much!

We did respect our teachers, even if they were very strict. We often got the cane or strap, in some instances for being late for school, which was hard on those who had a distance to travel. At other times perhaps we deserved the cane! The school was made up of three rooms. We began in the infants with Miss Gaskell and then moved into a class room, which is now the kitchen, where Mrs Barrow taught classes 1 and 2. After this we went into the big room which was L-shaped; it has been altered considerably since those days. We all went to school until we were 14 years old and the numbers were quite large, often 80-100, all from Great Langdale as Little Langdale had its own school.

The population of the valleys at this time was much larger with big families, in some cases eight or ten or more in each family. There was no such thing as school dinners and we had to go home for ours – some of the children who had longer distances to walk took their dinner in a basket. The school was dark as we had no electric light and it was sometimes so cold that we could not write. Then the teacher would send us out into the playground to run round and warm up. Dot Parker's mother used to make a delicious 'sugar shag' – bread and butter dipped in sugar! It was wonderful to relieve pain after an accident – after tumbling and scraping knees!

Accidents

On the whole the children were very well-behaved, although on one occasion, one of the boys had been into the quarry hut where the explosives were stored and had stolen some small brass caps or detonators. They looked very harmless and did not seem to contain anything more than a bit of dust. These caps had been distributed out as payment in games and most of the bigger boys had some of them. It so happened that the head teacher, Mr Barrow, was ailing and was off work and Mrs Barrow was teaching the whole school, apart from the infants. The girls were knitting and she had the boys reading and had them all in one big group. The boys, like most boys, were playing with these caps and even poking pins into them, when one exploded and blew Jim Haughey's hand off, blood spurted everywhere. He ran out of school and headed for home opposite the Glebe. Uncle Harry, who had the local garage, took Mr Hayes, the vicar, and Jim at top speed to the Westmorland County Hospital in Kendal. All the other children rushed out of the school, the girls carrying their knitting with them, but leaving the balls of wool behind – the result was a maze of tangled wool. When the boys realised what this cap had done, they went out and threw all the rest away. Later, it was quite a job finding them – one had actually been stepped on and crushed. Mr Barrow was furious and the children were all very frightened. This was in 1918. I was in school at the time but I hadn't procured any caps myself! Discipline was very harsh after this incident.

Jim eventually returned and was later fitted for a false hand which he kept concealed in a glove. Fortunately, it was his left hand so that he could write, ride a bicycle, and in due course drive a car and hold a pack of cards. He was errand boy for Fleming Mawson before obtaining a good job in insurance. He remained unmarried.

Games

The boys had their own playground and the girls had theirs with a separate door for each – we were not allowed to mix at all and we were not allowed to go into the wood which surrounded the playgrounds. We had various school games – marbles and checks were a favourite sport. There were coloured glassy marbles which were equal in value to about five of the plain ones. We used to be able to get bottles of lemonade with a marble in the top and we would break the bottle to get the marble out. These bottles are not seen nowadays. We always used to have a hoop or a 'booly', some of wood and some of steel, at certain times of the year. Most of these games were seasonal, like conkers in vogue in the autumn when the chestnuts were ripe. Skipping was another pleasant game – mostly with the girls. Bows and arrows were more popular with the boys. They usually made their own from wood, usually the yew; a catapult was very common too, but could be quite lethal. Spell and knur, and shuttlecock were two very interesting games. A 'bawtry' gun was made by taking the pith out of a piece of elderberry ('bawtry' was the dialect name for the elderberry). Most boys always carried a knife, a useful tool for whittling pieces of wood, sometimes it was used wrongly on school desks, but you would be in trouble if you were caught! Rounders or bat and ball were always good games, so too was leap-frog, and fox and hounds which was good if you had a lot of open ground or fell. All these games were very much enjoyed and were good exercise, too.

Lessons

In years gone by a slate was used for writing, but we used pencils at first, then when we were older we used pen and ink. We each had an individual ink well on our desk and these were filled using an old teapot. The pen nibs had to be renewed occasionally. You could produce a rich stylish type of writing with these. The modern 'biro' cannot give you the beautiful style we used to be able to get with the old type of pen. In school we had two extremely old long desks to seat about eight, and some others that could seat five or six – they were made of oak and were very heavy.

My period at school was during the First World War, and with the slump years which followed, we had rather a lean time. Our lessons were the basic reading, writing, and arithmetic, plus history, geography, and religious education. The bigger boys had gardening – there was a vegetable patch for 12 boys and each had his individual plot. I enjoyed this as I was very keen on plant life and growing things. The girls had sewing and knitting while we were doing this. We did quite a lot of singing and on occasions we gave concerts. I do not remember any tea-parties as those were very lean years. In the dark days of winter when it got dark very early our master would let us go home a bit earlier, that is the ones who had a distance to go, in order to get home before the light had gone completely. When the war finished we all learnt the various national anthems of the allies which we sang at the peace celebrations which were held in the Elterwater meadows. We had tea in the old Drill Hall after we had held the sports.

I have an extract from the parish magazine for January 1899:

> On Friday December 23rd the Vicar on behalf of the Managers presented Mary Stables with a silver watch and chain subscribed for them in recognition of the fact that she attended school for 8 years, and that during that period had never been absent and never late, making a total of 3,405 attendances.

This was my aunt who died of T.B. at the early age of 19 years. She was born in August 1884 and lived at Walthwaite Farm all her life. T.B. or consumption was a real scourge in those days and claimed the lives of many young folk. I still have her watch in my possession.

Leaving school – the grown-up world

I remember when I was a boy of 14 and had just left school, my father gave me about 20 hens so that I could make a bit of pocket money as we did not get a wage. One morning when I went to let them out, I found to my horror that a fox had been in and worried the lot, one was dead in the hen-house and he had bitten off the heads of all the others and he hadn't eaten one. I had forgotten to fasten them in properly – I had shut the door but I had not let the pop hole down. I was carrying some of the corpses away out of the fold where I kept them when I met my brother – he saw what had happened and laughed and said 'Bankrupt in one night!' I didn't think it was so funny myself – a quick end to my first business venture. I cannot imagine what those poor hens went through before the carnage finished. My father bought me some more and I was more careful to fasten them in at nights after that.

CHAPTER 10
LANGDALE CHURCH

Early chapels

For 300 years Langdale formed part of the parish of Grasmere and the dalesfolk attended the church there, occupying the north aisle which came to be known as the Langdale side. A field near the church was reserved for the horses of worshippers from Langdale, just as in later days the stable of the old White Lion at Chapel Stile was set aside. For at least 200 years Langdale returned two churchwardens and six sidesmen at the Grasmere Easter vestry meeting.

There was a chapel at Great or Mickle Langdale as early as 1571. There was also one in Little Langdale which stood near The Bield in a field called Chapelmire. There are no signs of this chapel today. The one in Great Langdale was small and most probably stood on or near the site of the present church in Chapel Stile. This chapel was kept in some sort of repair and was served, as were most of the Lake District chapels of this type, by an unordained 'reader' who would be an intelligent local man, but with no special education. It is interesting that the Langdale parishioners, and not the rector of Grasmere, presented him to the bishop. It was a very poor living and he was allowed by the bishop to sell ale to supplement his income. He also went from house to house for meals and stayed a certain number of days at each, using his 'whittle' or knife at the table with the family; this was known as 'whittlegate'. The names of a few of the early readers are known including 'William Gollinge of the Thrange in Langden, clerk' whose son was baptised at Grasmere in 1590. There was also a Charles Middlefell who died in 1643.

On Speed's map of Westmorland published in 1675 chapels are marked in both Langdales but in 1692 the Revd Thomas Machell of Kirkby Thore noted that the chapel in Little Langdale was long demolished. In Great Langdale he found the chapel 'in pretty good repair but for ye floore'. He drew a small sketch in the margin of his report. In 1693 the usual payment of 2s.6d. or 3s. continued to be made for the yearly 'rushbearing' since the first chapel did not have any floor-covering. Later the rushbearing hymn was written by the Revd Owen Lloyd, the curate at Langdale who was buried in Langdale churchyard. This hymn is still sung in both Ambleside and Grasmere at their rushbearing services:

Our fathers to the house of God,
As yet a building rude,
Bore offerings from the flowery sod,
And fragrant rushes strewed.

May we, their children, ne'er forget
The pious lesson given,
But honour still, together met,
The Lord of Earth and Heaven!

Sing we the good Creator's praise,
Who sends us sun and showers,
To cheer our hearts with fruitful days
And deck our world with flowers!

These of the great Redeemer's grace,
Bright emblems here are seen!
He makes to smile the desert place,
With flowers and rushes green!

Early in 1751 the old chapel was found to be very much in need of repair, so it was finally decided to rebuild it. During the next three years it was rebuilt at a cost of £100. This new chapel was paved so that there was no further need to bear the rushes or have the special service. Thomas Jackson, who became curate in Langdale in 1781 at the age of 23, was rector of Grasmere from 1806-21. However, he continued to hold the Langdale services himself as he owned Harry Place and lived there. Travelling to Grasmere on horseback from Langdale he decided that the Langdalians could well avoid the troublesome journey to Grasmere for funerals – a journey which involved the difficulties of the 'corpse road' over Hunting Stile. So in 1821 the little chapel yard was enlarged and the land consecrated. It was not until March 1828 that the first burial actually took place at Langdale, of one Margaret Harper. In 1849 a large extension was made to the original burial ground in which Owen Lloyd had planted the yew trees.

The chapel was not licenced for marriages until 1860 but this did not prevent various irregular marriages taking place, the first in 1835 being, curiously enough, between Adam and Eve Fleming, who occupied a minute cottage, now forming part of Thrang Cottage, owned by Hugh Parker. The first child to be baptised in Langdale was William, son of Henry and Mary Kirkby of Harry Place, in 1827 by the Revd Thomas Sewell. As the years went by this second chapel was in need of repair and the building was much too small being only some 60 x 28 feet.

The present church

The population rose from 297 in 1811 to 520 in 1851 and it was to rise still further, to 846 in another 50 years. The trouble was that a new church, though necessary, could not be afforded, for in 1845 most of the vicarage had been built at a cost of £400, divided between local subscriptions and Queen Anne's Bounty. Previously the curate had lived in what became the White Lion Inn, first mentioned as the parson's house in 1762. Now there came an 'off-comer' whose generosity, and that of his heirs, to Langdale can scarcely be exaggerated – Mr Edward Balme Wheatley (later Wheatley-Balme), a Yorkshireman from Mirfield who bought High Close. He had already built and endowed many churches and church schools in other parts of the country. In 1857 he and Mr John Robinson of Elterwater Hall applied for a faculty to rebuild the church and together they met the cost of nearly £1,600. The design was by John A. Cory in a Victorian version of the Decorated style. So rapid was the construction that on 2 July 1858 the bishop of Carlisle was able to dedicate the present church to the Holy Trinity. It was enlarged in 1879 by E.B.W. Balme as it had proved to be too small for the needs of the parish.

In 1863 Langdale was separated from Grasmere for all ecclesiastical purposes and became independent under the new Parishes Act. The vicarage was altered and improved again in 1884 through the generosity of E.B.W. Balme. The minister's house was sold before the building of the new vicarage and became the White Lion Inn about 1843.

Many gifts have been made to the church over the years. The peal of six bells was the gift of E.B.W. Balme – they were cast in the Whitechapel Foundry in London by Messrs C. & G. Mears and are remarkable for their sweetness of tone. The clock was made by Harrison of Hull who was a noted turret clockmaker of his day. He completed the mechanism in 1858, but refused to install it in the tower until he had tested it for five years in all conditions of heat and cold.

In 1891 some pews were removed from the north aisle to allow W. Rushworth & Draper of Liverpool to install an organ which was given by E.B.W. Balme and Albert Fleming. The first time the organ was played at a wedding was that of my parents in 1896 – John Fletcher Buntin and Hannah Stables. Mr Fisher was the organist for many years followed by Miss Lawson, then Mr C. Sandford. After this my sister Hannah Fletcher Buntin played for the remarkable time of 65 years from 1919 until she retired in 1984.

The church was first lighted for every service in 1887. The parishioners subscribed about £25 to purchase 12 lamps supplied by Wright & Butcher of Birmingham. More recently in 1956 when electric power came up the valley, it was installed in the church at a cost of £700. The embroidered hangings of the pulpit, lectern, and reading desk were the work of Miss

Hannah Robinson. The heating of the church which was produced by a coke-fired boiler in the cellar got into a bad state of repair so was replaced by an oil-fired boiler.

Colonel Robinson of the Gunpowder Company gave land for a new cemetery in 1915 as the churchyard was almost full – this was consecrated by Bishop Diggles and the wrought-iron gates were erected in 1951 from subscriptions, Mrs Mawson being the chief instigator.

The parish magazine in 1901 recorded a very special gift: a very beautiful and valuable 'fine linen cloth' was presented to the vicar by Mrs Pepper of St Martin's, as a parting gift on her removal from the 'Spinning House' to Tilberthwaite. It is made from the best Langdale Linen and the needlework is of the Flemish pattern. The cloth was used for the first time on St Martin's Day 1901 at a celebration of the Holy Communion and still exists today.

Readers, curates, and vicars of Langdale

John Hunter	1554
Thomas Benson	1563
William Gollinge	1590
Charles Middlefell died	1643
Richard Harrison died	1670
James Clark, last reader	1728-43
Richard Steele	1743-80
William Crawford	1780-81
Thomas Jackson, assistant	1780
Thomas Jackson	1781-1819
William Jackson	1819-28
Thomas Sewell, assistant	1820
Daniel Green	1828-29
Owen Lloyd	1829-41
Haygarth Barnes, assistant	1832
Stephen Birkett	1842-60
James Coward	1860-85
R.S. Hulbert, assistant	1883
Robert S. Hulbert	1885-1905
John M. Tredennick	1905-11
William Hayes	1911-25
Leonard Argyle	1926-30
George C. Dunning	1931-53
Geoffrey W. Ellison	1953-61
Richard R. Saville	1961-72
Graham Hartley	1972-81
Derek G. Hall	1981-86

Walter Wade	1987-
Delmar Banner, lay reader	1956-83

The church today

The church in the Langdales continues to prosper with services at Skelwith Bridge and Little Langdale as well as the main services at Chapel Stile. These cater both for locals and for visitors from all over the world. Family services are lively and coffee afterwards gives an opportunity to meet informally and to welcome new members. Parish walks and picnics after church on Sundays, even in December, and on Bank Holidays have proved a great success. A new Explorer Group and the Youth Fellowship help to involve the young people who have enjoyed various outings and their own camps at Blea Tarn.

The old fireplace at Birk Howe

CHAPTER 11

GREAT LANGDALE FARMS

Great Langdale was part of the Barony of Kendal, each farm paying small customary rents to the owners of the Barony, who had the manorial rights. In the last century the lord of the manor was the Earl of Lonsdale. When a holding passed from father to son a 'fine' was paid by the new owner, and also when the lord of the manor died.

Baysbrown, or Basebrun

This is one of the oldest and most important holdings in Langdale. Records show it to be a separate small manor which comprised the north-east corner of Lingmoor, from the watershed down to the river which runs down the valley, and onwards to include most parts of Elterwater. In the reign of Henry II (1154-89) it belonged to William of Lancaster, Baron of Kendal. It was in this reign that Conishead Priory near Ulverston was founded and William of Lancaster not only made a grant of the land on which it stood but also endowed it. Amongst the endowments was the manor of Baysbrown. After the Reformation it passed through various families including Benson, Braithwaite, Atkinson, and Harrison until in 1919 it was sold to Mr Pattinson. The rent was then £115 per annum.

The present house, traditionally built, is dated 1687. The porch is unusual being extended on one side; the house is large with seven bedrooms and has a beautiful court cupboard. It has extensive farm buildings and land now including that from Oak Howe which used to be a separate holding. The house and buildings are situated on the sunless side of the valley and for quite a long period in the winter the sun never shines on them. It is approached either by the lane off the main road as you leave Chapel Stile or by the Ullets (Owlets) Nest road. It has a large amount of surrounding woodland, now a conservation area, and numerous slate quarries on the Great Langdale side of Lingmoor.

The farm has one stock of sheep running on Lingmoor and one on Bowfell. Early records show 631 plus 126 wethers. The sheep marks were: Lingmoor stock, cropped both ears; smit – pop on tail head; Bowfell stock, cropped both ears; smit – a pop on tail head and one on top of fillets and a pop on top of the head, with an 'H' on the nearside.

Occupants were Thomas Braithwaite 1653, Gawen Braithwaite 1664, Thomas Borwick 1664, Joseph Workman 1817, R. Greenup and D. Dawson 1837, William Greenup 1871, William Gregg and John Allonby 1879, William Jackson, 1922 (only there for one year and died rather tragically), William Russell 1923, Joseph Gregg (son of William Gregg) 1926. In 1932 George Grisedale and his son, John, who carried on after his father's death until he retired on 12 November 1971; Noel Hodgson 1972 till he died in 1988, David Rowand 1989.

Field names: Fitz Head, Back of Beck (Bacabeck), Broad Meadow, Straight Meadow, Gala Field, Calf Parrock

Oak Howe, or Yak Howe

This was originally a very ancient holding standing on the sunless side of the valley under Lingmoor. It consists of quite a small old house made of local slate together with the adjoining farm buildings. The buildings do not get any sun for some time during the winter. Tradition connects this farm with victims of a pestilence as mentioned earlier.

Occupants were Anthony Dixon 1578, Robert Hawkrigg 1726, Jacob Fletcher, a tailor, 1754, Joseph Workman and E. Parke 1817; Oak Howe has been occupied by various tenants in recent years.

About 1817 the sheep marks were cropped near ear and underhalved far; smit – a red pop on the near lisk and a pop on the tail head; wethers smit on the head, twinters a pop on the shoulders.

Harry Place

Many farms are named after the people who lived there, it is most probable that this one was named after the Harrisons in the 1600s. It is another old farm holding but showing different architectural features as it has rounded chimneys and pointed windows. Originally it had five bedrooms, four on the front and one at the back but this has been recently made into a bathroom. The farm is supplied with a good range of buildings. Perched on a raised piece of ground the situation is beautiful with fine views right up the valley to the Pikes, Crinkle Crags, and Bowfell. However, with the present volume of traffic, it is rather too close to the main road and is very dangerous for the cattle, sheep, and particularly dogs.

Early records show a stock of 158 sheep. The mark was forked both ears and the smit mark was an inverted 'V' on the nearside. A while back the sheep were given an 'I' face burn but this was never very good so that it was finally left off.

Field names: Mirk Holme, Acre Dales, Miller Garth, Crook Holme and Kepple Garth, By Croft, Little Croft, Wood Hinning and Corn Hinning, Big and Little Robsay, Galefitts, Matthew Meadow, Hen Parrock, Strait Intake, Willy Wife Intake and Plantation, Stony Close and Parrock.

On Low Far Intake, a house called the Studio was built in 1931 by B.D. Walmesley in which he produced his lovely stained glass windows. On Thrang Close a house by that name was built in 1934 by Mr Bulcock, and on Copt Howe and the Plantation a house called Copt Howe was built in 1933 by Miss K. Ackerby. Rutts, a field across the beck near Oak Howe was sold off to Baysbrown, Harry Field in Chapel Stile, at the top of the common, was sold to J.M. Thompson, and Broad Howe was sold to Mr Balme, but this has recently been put back to Harry Place. These three pieces were sold off by Tommy Greenhow who owned the farm at the time.

William and James Harrison occupied the farm in 1684, James Dixon in 1817. The Revd Thomas Jackson owned it soon after, lived there and died there in 1822 – he used to ride his pony most days over the fells to Grasmere where he was rector. The first recorded baptism in Great Langdale church was of William, son of Henry and Mary Kirkby of Harry Place on 1 July 1827. Other occupants were Andrew Rigg 1851, Edward Philipson 1860-89, Casson Stevenson 1889-98. Tommy Greenhow was tenant for George Cragg who bought Harry Place in 1897. In 1920 he bought the farm and died there in 1932. He married Anne Agnes Harrison in 1880 and they had a family of eleven, four boys and seven girls. After the death of both parents the farm was worked by the family until 1964 when Mr and Mrs Phillip Powell came to farm it until 1973 when they left to farm Brimmer Head in Grasmere and Mr and Mrs C. Benson took their place. Professor G.M. Trevelyan bought Harry Place from Tommy Greenhow's widow and gave it to the National Trust in 1944.

Harrison Place, now Robin Ghyll

This was most probably a small farm in the early days. The original house was quite small but it has been extended a few times. It is in a beautiful situation with commanding views of the valley.

In 1692 Anthony Harrison, cordwainer (shoemaker) was in business here. I have a document referring to the transfer of the field called George's Tenement or Shop Field to Thomas Fearon of Robinson Place by Mr Harrison in 1693. In 1799 it was occupied by John Greenup who had a stock of 150 sheep. About 1905 it was let to Miss Dorothy Ward, daughter of Mrs Humphrey Ward the novelist, and Professor G.M. Trevelyan, her brother-in-law, used to come here many times on holiday. He was a great lover of Langdale and bought and gave a number of farms to the National Trust in his latter years including the present Millbeck Farm and Harry Place. He bought Robin Ghyll from Mrs Cragg in 1911 and it was bought by the National Trust in 1979 from his grandchildren.

Robinson Place

Robinson Place is one of the old farmsteads of Langdale, situated about half way between Chapel Stile and Dungeon Ghyll. It has the initials and date on the front of the house T.F.A. 1693. It has been extended and renovated many times but fortunately the plaque with the date is still intact. The character of the house has sadly been lost to some extent with the alterations. There are two beautiful old spice cupboards inside the house and one is dated T.F.A. 1692 which is earlier than the date on the front of the house. This shows that the date on the house is not necessarily the year that the house was built, but rather that it was renovated then and the owner put his initials on the front. In this case T.F.A. stands for Thomas and Agnes Fearon. In the baptism register of Grasmere church there is a record which refers to a son of Thomas and Agnes Fearon of Robinson Place, 1693.

Inside the house there is a beautiful old oak partition in one of the bedrooms and one very old black oak door. During the last alterations in 1925 it was discovered that a few windows had been blocked up; this was no doubt to cut down costs when the window tax came into being. A doorway had also been blocked up showing that access had been possible from the house to the attached shippon, avoiding the need to go outside. Some of the walls are about 3 feet thick and were made to stand for hundreds of years.

The view from Robinson Place is one of the best in Langdale. The towering Langdale Pikes, Bowfell and Crinkle Crags and Pike o' Blisco are just magnificent and are easily climbed in no more than a few hours. Even the comparatively small Lingmoor is a marvellous walk and from the top you have a commanding view of all the farmsteads in the valley. The garden too is a sight to see and has been created by the Buntin family. It is a very sheltered garden standing under the eastern side of the fells and extends to about half an acre. There are the usual yew trees which were planted round the old farms and not far from the house are three huge Scots pines. My grandmother used to say that there were 'Seven Sisters' but now only three are left. Rhododendrons and azaleas are a lovely sight in the spring; there is also a good collection of heathers which love the acid soil. With Robin Ghyll Beck running through the garden it really is a marvellous sight and has been photographed many times.

Early records show that the stock of sheep went on Bowfell in the summer and in the winter they were turned out on Silver Howe and Raw Breast Fell. The smit mark was a stroke over fillets and a pop on the tail head; the ear mark was forked near ear. This stock eventually left Bowfell and was run on the home fell; this was in 1924 or 1925. John Buntin had already bought Pye Howe with a small stock of sheep – the marks being a stroke down the far shoulder and another stroke on the far lisk, with ear marks, forked both ears and punch holed far ear. The old Bowfell mark was done away with and the Pye Howe mark used from that time. It is interesting to note that in an early *Shepherd's Guide* the Bowfell stock was named the Ellers stock by Mr Youdell; obviously he must have kept the sheep there when he went to live at Ellers, after he sold Robinson Place to John Buntin. Langdale was 'unstinted', you could keep as many sheep as you wanted, but with the 1964 Commons Registration Act everyone had to register his sheep. We then registered 1,300 sheep and followers.

Field names: Back Meadow, More, Dowy, Long Lands, Hard Butt, Greenmire, Causeway, Helm Howe, Breeches or George Close, George's Tenement or Shop Field, Thorny Intake, Low Pye Howe, High Pye Howe, Wormy Crag Intake, White Crag Intake, Across Ghyll, Behind House, Behind Middle Barn, Behind Ellers, Under House.

I have several documents relating to Robinson Place. Thomas and Agnes Fearon were here in 1692. It was sold to John Youdell before 1747. In 1774 Anne Youdell relinquished her widow's right at Robinson Place and lands at Stonethwaite in Borrowdale in favour of her son John, on the payment of 5s. He died in 1822 and was followed by his son Joseph (d. 1829) and then the farm was divided between Joseph's sons John and Joseph. John bought out his brother in 1832 for £480, raising a mortgage of £500 from Thomas Coward, yeoman of Colt Park, and James Ashburner of Bardsea, maltster.

A small stretch of water above Robinson Place, found by ascending Robin Ghyll, is known locally as Youdell Tarn. It was probably formed by past peat-digging. A large colony of black-headed gulls nest here now. About 300 yards past the tarn in an easterly direction, following the path under Lang Howe and over the rise, you drop down to Geordie (George) Walker's Peat Pot. We called the path 'Langwray'. Little is known of Geordie Walker but on the back of the barn door at Robinson Place is painted in red 'Geordie Walker – demon bowler of Langdale Cricket Club'.

John Youdell died on 9 April 1874. When his son John, nicknamed 'Gentleman John', left to live at Ellers, the farm was tenanted by Thomas Atkinson. The Youdells owned this farm until it was sold on 17 February 1883 to my grandfather John Buntin.

John Buntin died at Robinson Place on 10 August 1910 aged 78, his father also died there on 8 July 1886 aged 89, and my father also died there on 6 February 1969, so did his eldest son John Fletcher on 20 January 1960. Four generations of John Buntins died there. It was farmed by the Buntin family until 1974 when it was sold to the National Trust.

Ellers

The name 'Ellers' is the local name for the alder trees. Ellers stood on the site of the present Ellergarth. It was a very old and crudely constructed

Hugh Parker and John Mitchell laying a hedge under the Pikes.

Vic Gregg winter-feeding Herdwicks at Low Millbeck.

'Storm Clouds over Langdale': the Buntin family leading hay at Robinson Place in the 1930s.

building with a barn attached to the end. No doubt it had been a small farm originally and was approached from the lane which leads up to Robinson Place. The shippon with a small barn at the end was derelict so we made it into a garage. In front of the house, on the opposite side of the lane was a blacksmith's shop and a joiner's shop, in the field called Shop Field. These were pulled down and the stone used to build the new house of Ellergarth but you can still see traces of the old hearth stone. On the north side of the house was a hoggus with a small vegetable garden; this still exists and was put to Robinson Place. Ellers was given to my mother and father as a wedding present, and they lived there for a year or two after being married in 1896. Their eldest child, John Fletcher Buntin, was born there in 1897.

It was in the barn next to the house where my father was chopping wood one night, using a candle for light. It was an old draughty building and when the wind blew the candle out he chopped off his left thumb (and he could never milk after that). He ran about frantic, his first place of call was his parents who lived just up the lane at Robinson Place. They had no idea about stopping the bleeding and his mother tried cobwebs. Eventually his father yoked the horse and trap up and conveyed him to the doctor at Ambleside, who stopped the bleeding but father had to stay in Ambleside to be near the doctor. He stayed a few days with Mrs Dugdale in Church Street – she sold 'claggan' in her toffee shop – a delicious sticky toffee. They were afraid that my father might have lockjaw setting in, the finger next to the thumb was almost severed too, but they managed to save that. I think they buried the thumb at Robinson Place!

So it seems that this old holding stood here for a very long time and has a long history of occupiers. My grandfather, John Stables, told me that he went to school there and was taught by Mr Spedding, 'Auld Will Speddy' as he called him, and that he was a very learned man and could teach Greek and Latin. He also told me that he once barked some oak wood in the joiner's shop. This was used for tanning leather; whether this bark was for sale I do not know but as my grandfather was a shoemaker he would be very interested in tanning the leather.

Old John Youdell died at Ellers in 1888; he must have gone to live there after he sold Robinson Place to John Buntin. My father and mother left Ellers in 1898 to go to live in Pye Howe and after that it was occupied by various people until in 1909 my father sold it to the Revd William Hutchinson Sharpe, a former vicar of Isel near Cockermouth who had lived at Mere Syke. Ellers was dismantled by him and Ellergarth was built on the same site, Jonathan Stables being the architect. Mr Sharpe's first wife was a titled lady from Isel Hall, she died with no family; to his second wife he had a son William born in 1900, Charles in 1902, and Mary in 1905. Unfortunately he had an accident and never recovered properly and died before the new house was completed. His widow and three children lived in Ellergarth and they had a governess, Constance Eltis; Mrs Sharpe died in 1914. Ellergarth has been occupied by a number of people, the Hon. Mrs Meade, Mr and Mrs Seabrooke. Mary Sharpe and Constance Eltis came back for a short time in 1924 and were followed by Mr and Mrs Colin B. Phillips. An artist was there for a year or two; finally Mary came of age in 1926 and Ellergarth was up for sale. It was sold for £1,500 to Professor Cyril Montague Sleeman, Professor of Physics at Queens' College, Cambridge. He and his wife lived there until their deaths and it has now been bought by Mr and Mrs Meschino.

Johnson Place

Very little is known about this small holding, it was probably in the vicinity of Ellers. It appears in one of my documents and in a list of the Kendal Barony rents in 1689, where it is stated that 'William Tyson died seized of a customary tenement and messuage at Will Johnson Place'.

Pye Howe

I was born here on 15 February 1910. It is a real old-world farmhouse, now a cottage. The origin of the name is uncertain. Some have suggested it is from the magpie which is certainly common in the vicinity now. It used to be dirty white in colour but it has been pink-washed in recent years, otherwise it has not been altered to any great extent, only a little corner end has been added for a bathroom. An interesting feature is the tall chimney from the old kitchen. The front door opens into the living room where there is a lovely spice cupboard to the right of the fireplace; a parlour is situated to the left as you enter. The ground floors are of blue stone and the old partitions are made from beautiful oak with black oak doors. The stairs are of stone halfway and at the turn they are made of oak which squeaks with every step you take – not much good to any one of the family coming in late – your parents were sure to hear! There are three good-sized bedrooms, one with a lovely old fire-grate, almost like an Adam grate.

Outside, next to the house, is the barn or peat house which is of very rugged construction and had the old wrestler slates for ridging, now replaced by modern ridging tiles. The garden is well kept and has the usual yew trees, and the view is magnificent. In the wall to the west of the house is the bee house which would have taken two bee skeps in the old days. My father was a very keen bee-keeper and he had about a dozen hives while he was at Pye Howe which he gave to me when I left school in 1924. The early honey was thin and watery, but in a fine summer the bees yielded an early spring flower honey from the sycamore, hawthorn, lime, and clover, followed by the late heather honey in August. This was thick and viscous –

there is nothing like comb honey. He didn't have an extractor so he always made comb or section honey. A section would yield a pound of honey, in those days it was worth about 1s.6d., now about £2. The straw skeps were used before my father's time, with the hole in the top fitted with a wooden bung during the winter; in the summer the bung was removed and the hive was capped with a straw cap which was where the bees stored the honey. We used to use the straw skeps only to collect a swarm.

Pye Howe was a small farm when John Buntin, my grandfather, bought it in 1896 for £800. It had a small stock of 80 Herdwick sheep. The mark was two strokes down the far side and forked both ears and punch holed near ear. John Fletcher Buntin, his only son and my father, moved from Ellers to Pye Howe in 1898. In the year 1900 my father had a difference with his father and mother and he took off to America. My mother was left behind and went to live with her parents at Walthwaite for a short time until she moved to a cottage at Lingmoor Row. I think his intention was to have his wife follow him to the U.S.A. If this had happened the rest of us would have been born in America. However, a reconciliation happened and he returned to live at Pye Howe again where the rest of the family were born and he worked for his father again. His father died in 1910 and my grandmother lived on at Robinson Place alone. Eventually, we all went to live at Robinson Place in 1923, grandmother having had a stroke in 1921. It was difficult keeping two homes going. Another reason for moving was to help a Mrs Jackson to move in as her husband had come to a tragic death at Baysbrown. They had to get out of Baysbrown immediately and they had a large young family and were not very well off. We adopted one of the girls, Elsie who was aged seven, who now lives in New Zealand. My grandmother, Jane Buntin, died in 1925, nearly 90 years old.

One early occupant of Pye Howe was Robert Harrison, whose son Thomas was baptised from the house in 1707. In 1870 William and Agnes Thompson lived there, and they were followed by John and Agnes Pattinson in 1871 and later a Mr and Mrs Graves, also a Mr and Mrs Birkett, and the Martindales were there for short periods. Pye Howe was owned in 1893 by a Mr Schollick, whose wife was the grandaughter of the Revd Thomas Jackson, and he also owned Harry Place. He tried to sell with no success in that year so he sold Dale Field then and eventually Pye Howe in 1896

Field names: Low Becks, New Field, Near Cunsay, Mirk Holme, Stillings Foot, Long Leg and High Ford, Drum and Low Ford, George Close, Old Garth and Wet Parrock (where Heron Crag and Heron Place were built in the 1930s), Thorny Parrock, and Croft, Newlands, High Pye Howe, Thistly Parrock, George Intake, Old Intake, Big Bitts, Robin Intake, New Intake, Long-tailed Cross, Far Cross, and Dale or Town Field which was sold off and put to Raw Head.

A good few years ago a Lady Simpkin lived by herself and made the garden of Heron Crag; she kept two spaniel dogs. She thought the world of these two dogs. They were fed off all the best food – steak and the like – they used to stand on a rock in the garden and bark the whole day through. This annoyed my father very much. When the dogs died she had beautiful coffins made for each one and they were buried by the fence in the garden. Around the two graves were kerbstones and a tombstone with their names inscribed. When Lady Simpkin died she left the little bit of ground where the graves were to my father, so the fence was adjusted to accomodate this. My father was very annoyed and said he wanted no dogs' tombstones on his land and vowed he would go and break them up. He was an old man by this time in 1961 – he was 87 years old then. I was not sure that he could get that distance and I did not like his attitude, so I went and took off the turf and laid the tombstones flat and covered them with the turf. It was rather an underhand thing to do, but I felt that I was doing the right thing. I do not know whether my father ever went to do the job but he would find that the tombstones had gone. The tombstones are still there, and maybe in years to come some one will dig them up and find them!

Bowderston, or Borderston

Between Pye Howe and Long House there was a small holding called Bowderston, later Borderston – we called it Borderson. Apart from a few walls and some piles of stones there is little evidence of what was on the site. This tenement was occupied by Alexander Tyson and John Dickenson in 1573-74 when each had to pay a yearly rent of 2s.2½d. to Henry, Earl of Cumberland. A century later it was called Borderston and had been in the possession of John Grigg, John Benson, and Thomas Borwick, but now it was James Dixon who had to pay the yearly rent of 1s.

Long House

When Long House came up for auction sale on 12 September 1898 it was described on the auctioneer's advertisement as

> . . . comprising Dwelling House, Out Buildings and Land containing altogether 22a 2r 29p in the occupation of James Grave. The farm is pleasantly situate in the picturesque valley of Great Langdale. The Lakes and rivers in the vicinity afford excellent fishing. The Dwelling House is commodious and the Barn New . . . with the property is enjoyed unlimited rights on the Commons of the Parish. The Purchaser of this lot is to take the Stock of 70 Herdwick Sheep, let with the Farm for which he shall pay £45 in addition to his purchase money.

This was once a little farm holding. The house is the original with a small addition on the south end for a bathroom. The barn, now converted for holiday use, is a comparatively new one and very well built. It stands in a beautiful setting with an attractive garden; in the past the orchard yielded a lot of fruit. The usual yew trees stand in front of the house. A hundred years ago this small farm had not many fields, just Garden Field, Stillings, and Back Meadow, with the intakes on the higher ground behind. Joseph Youdell owned High and Low Mucktin and Borderson and when he bought Long House it was all made into one unit.

An old walled outgang starts just behind the house, it aims to go up and over the fell by Pye Howe sled gate. By some people this was claimed to be the route and a case was brought up about 20 years ago claiming the right of way as passing the house. This case was fought by Mrs Satow who was the owner and lived there. Naturally she did not want this right of way which would spoil her privacy. She won her case. I was there at the hearing in Ambleside. Afterwards she bought the intake called Pye Howe and a little piece further down the road, this at one time belonged to Raw Head and had been sold off separately when John Robinson's estate was wound up. This intake goes from the main road right to the fell wall. She then gave a right of way up here and made a stile at the bottom and one at the top. This is a good route and has joined up with the Pye Howe sled gate at the top. This intake divided the Pye Howe and Long House intakes. There was a need for some such path to give access to the fells – it was a good move. She then gave this intake to the National Trust.

I cannot go very far back with the people living here. Robert and Hannah Greenup were there in 1871 and were followed by Joseph Youdell. Thomas Tyson took over in April 1885 but he died shortly after – he was only 31 years old. Then Thomas Greenhow moved here from Wilson Place in April 1887; he left in May 1895 and went to live at Raw Head Cottage for a time and James Graves took over. He stayed until 1913 when William Youdell (Dot Parker's uncle) took over but he only stayed for a year.

In April 1914 William Parker, Dot's father, took over and stayed until 1918. He was known as 'Morc' as he was born on the Morecambe Trip Day, and worked as my father's farm man and married Louisa (Weety) Youdell. He had a great sense of humour and loved to tease people and was always up to pranks. At one time he was very annoyed by a girl taking her pram across his long mowing grass and leaving it on the rocky hillock, Helm Howe. While she was away talking to the workers at Ellergarth he wheeled the pram down and put it into the soft bog and then awaited the result. The girl eventually returned and found the pram, but in retrieving it she slipped into the soft ground, right above her knees. She never went there again! William was in Long House during the very hard winter of 1918 when several of the sheep were trapped under the snow. One Herdwick survived being under the snow for 35 days. He left to go to Elterwater Farm.

Old Joseph Youdell came and stayed for a year or two with his second wife and her brother, Joseph Wardley. Finally it was sold in 1924 to Percy Clark and it was no longer run as a farm. My father took the land and had it for a few years. The next occupants were Samuel Briggs and then Mr Valardy followed by Mr and Mrs Satow, Mr and Mrs Potter, and now Mr and Mrs Wilkinson. Now the bottom land is owned by Mrs Satow junior and the intakes by Mr Holden. The lower land was owned by Mr Blackburn who farmed Side House along with Raw Head fields; when he sold up Mrs Satow bought both these two pieces of land and this ended the old farm holding.

There was a small stock of 70 sheep and the mark was ritted near ear and cropped far ear, and the smit marks were a crupper and pop nearside and a letter 'T' on the far side.

Raw Head

This farm was situated at the end of the old lane called Raw Lane or Raw Lonning and gets its name from being the head of the row of houses. The present whitewashed building is in an elevated position in the shape of an 'L', built in the usual style with a little piece added on to the side for a bathroom.

Inside the house are some really beautiful black doors, with a marvellous court cupboard dated T.H.H. 1689. Between the fireplace and the small fire window is cupboard space reaching from the floor to the ceiling, comprised of a double-sized spice cupboard above with a smaller cupboard below and a further double cupboard reaching the floor. In a panel below the middle cupboard, about two feet from the ground, there is a sliding piece of wood, like a moveable shelf 3 inches wide, which can be pulled out to about 18 inches and retains its horizontal position. Its purpose is a mystery – was it to support a candle or a small drop-down desk or was it for measuring materials? The flagged floors are still in existence, but a modern fireplace has replaced the old oven and boiler. The stairs are shallow and wide and are in good order and lead up to the landing which forms a balcony overlooking the stairs and access to the three large bedrooms which are furnished with bunk beds giving accomodation for 16 people. The house is beautifully preserved inside and I think the Fell and Rock have done a good job since they bought the house in 1946 – on the whole it is very much as it was a few hundred years ago.

Set into the front of the building are some attractive bee boles formed with slate slabs, and others are in the wall just below and to the west of the house. The usual yew trees are present and are particularly fine specimens. Scale Ghyll runs past the house supplying unlimited water.

At the back of the farm is a good outgang, which links with the Raw Head sled road, used for gathering the sheep and bringing down the peats.

The track, which is cleverly embanked in places, winds its way up the steep fellside over a lovely stone bridge to a little plateau under Black Crag. Here are the remains of six peat houses. If you follow the main track up to the north-west you come to several well-worked peat diggings, now filled with water. Still further on, between White Ghyll Crag and Blea Rigg, is a solidly built sheep fold, much better constructed than most. Another feature behind White Ghyll Crag is the remains of an old wall which was built to prevent the cattle from straying near the dangerous crags.

The old barn next to the house and to the east was bought by the Fell and Rock in 1947 from C.G. Bulman and has been made into an annexe – a very spacious hostel with a large sitting room and a steep ladder up to the bedrooms providing sleeping accommodation for 31 people. A plaque on the wall reads 'In 1947 the freehold of this barn was purchased by members of the London Section of the Fell and Rock and their friends in memory of G.R. Speaker'.

The large barn to the west of the house has been developed into a Roman Catholic hostel called Bishop's Scale and the hoggus about 100 yards up the valley by the roadside has been made into an attractive chapel. In a way it is a shame that a good little farm holding was broken up. The intakes were sold to the Roman Catholic Bishop of Lancaster, the Rt Revd Thomas B. Pearson. It was reported in the *Westmorland Gazette* for 13 August 1955 that he had bought a mountain side in Great Langdale! The mountain side which goes up 1,250 feet includes crag, farm buildings, 20 acres of farm land and several hundred acres of fell. This farm building was to be converted to give a permanent home to the Achille Ratti Climbing Club and was to accommodate 150 people when it was completed. This report was partly a myth as no-one is allowed to buy common land, but the intakes could be bought and the barn and shippon were converted into the climbing hut. The fields were sold and are now owned by Mrs Satow junior.

Field names: Big Ridding, Small Ridding, Long Thorn, High and Low Cottage Meadows, Croft, Threaplands, Codling Field, and Sandy Field which was given to Thomas Earl for work done. This field was owned by John Robinson and it was a liability as the beck fence had broken down and allowed the field to be covered with sand. He tried in vain to get the river authority to do the work but the outcome was the work was delayed for some months during which the river was running through playing great havoc. He decided to get rid of it – a wise move.

Raw Head was the largest of the holdings along the row and originally had a stock of 350 sheep but one tenant went bankrupt and the sheep dwindled to 124. The mark was cropped near ear and ritted far. The smit marks were a stroke down far shoulder and one close behind shoulder and a letter 'R' on the near side.

Occupants were Thomas Tyson 1673, John Johnson, Robert Harrison 1697, William Harrison (Robert's son) 1699, George Hodgson 1817, Isaac Tyson 1871, Nixon Armstrong 1886 (he died in 1887, aged 50), John Harrison 1888, Isaac Jenkinson 1892, James Andrew Dickinson 1904 (left and sold up in 1919), Fred Greenhow 1919 (left in 1931 and went to Hall Dunnerdale). Joseph Park came in 1931, followed by his son William until he left in 1944 and went to The Ellers in Loughrigg and Raw Head was finished as a farm. The land after this was farmed by C.G. Bulman.

Whitegill

Between Raw Head and Low Millbeck was a small tenement called Whitegill, just to the east of White Ghyll Beck where a hogg house stands today. The 1573-74 rental and survey of lands leased to Henry, Earl of Cumberland records the occupants at that time and the rent due:

Hugh Cradall	tenant	12d.
John Cradall	tenant	18½d.
Edward Cradall	tenant	18½d.

In a document in my possession dated 9 June 1576 John Yewdale received £19 from Edward Benson for half of the tenantright in the farmhold at Whitegill. Presumably at the same time he was selling the other half to someone else, now unknown. In another document dated 1734 Agnes Benson of Millbeck enters into an agreement with her son John, in which she gave up her right of dower at Whitegill and in land at Mickle Ing in Great Langdale common field.

By 1819 it is recorded that Edward Benson of Millbeck had to pay a rent of 8d for 'closes and parcels called High Close, Low Close, High Intack, Low Intack with a hoghouse, standing and being at Whitegill being parcel of a tenement at Raw', so it seems as though Whitegill had ceased to be a farm by that time.

Low Millbeck

This farm stands at the bottom of Mill Ghyll or Stickle Ghyll as it is sometimes called, as it runs out of Stickle Tarn. There was a mill here, possibly as early as 1283 although the precise position is not known. Not much is known about it but it was probably a corn mill and maybe a fulling mill was here in later years. It was long owned by the Benson family who were scattered about Langdale and were probably one of the oldest families in the district. High above Millbeck stands Miller Crag.

The earliest document I have relating to Millbeck is dated 1621 and concerns a dispute between Edward Benson of High Close and Robert Nicholson of Millbeck over the title to a tenement there. It was agreed that it did belong to Nicholson, but he had to pay Edward Benson 20 marks. In a deed of exchange relating to Millbeck, dated 31 May 1673, John Nicholson is moving to Elterwater and exchanges Millbeck for Edward Benson's property there. Edward Benson is to pay John Nicholson's debts, which amounted to £70. It is interesting to note that 'all lofts, jeasts (joists), loose boards, and stangs (posts) shall remain in the houses as now they do'.

In a conveyance dated 28 January 1699 John Benson of Millbeck, yeoman, paid Leonard Benson of Elterwater, yeoman, the sum of £14.5s. for a

> messuage and Dwelling house-stead at Millbeck . . . aforesaid; And alsoe One Close of Arrable land . . . commonly called Gooddy-Close; And alsoe one close of Pasture commonly . . . called Old-Close; And alsoe One close of woods and wast-Ground commonly called Lile-hagg; And one Dale of Arrable Land . . . commonly called Leas . . . And alsoe all his wood, wood-meer and Trees.

1714 In the window tax assessment, John Benson pays tax on nine windows.

1734 Agnes Benson, widow of John, gives up Millbeck in favour of her son Edward.

1767 Eleanor Benson and her son John mortgage Millbeck to John Gillbanks of Middlefell Place.

1774 Land tax assessment, John Benson has to pay 6s.3d.

1783 John Benson of Millbeck mortgages it to James Harrison of Monk Coniston.

1809 John Benson makes his will leaving Millbeck to his son Edward.

1871 James Hodgson living at Millbeck with 5 children.

1879 Thomas Birkett was at Millbeck; one of his three sons, Thomas Morris married Anne Chapman in 1901 and moved to Side House. The younger sons were Jonathan and William.

1884 Edward Benson (1772-1860), the last of this family, left Low Millbeck to his nephew, Robert Rollinson or Rallinshaw, son of his sister Eleanor. He died in 1884 and Low Millbeck was left in trust for his nephews, John and Joseph Plaskett of Stonethwaite in Borrowdale.

1907 Thomas and William Youdall, brothers, were tenants until 1922.

1921-36 Henry Mounsey farmed it. He sold it to Professor G.M. Trevelyan in 1932 who in turn gave it to the National Trust in 1944.

1936-85 Farmed by Joseph Gregg, then by his son Vic.

1985 Farmed by Eric Taylforth.

Vic Gregg recalls farming at Low Millbeck:

> We only ploughed during the war, and that was with the horse plough. We got terrific crops of corn off Barn Field, Leys and Mickle Ing. We planted in rotation yearly – corn followed by turnips and potatoes, then corn undersown with grass. Each time you

ploughed the stones kept coming up – so as soon as the war was finished we soon gave up ploughing. There aren't many hedges as the horses would eat them – they barked them – and they soon died. We used the sammel pit between White Ghyll and Raw Head but it was very dusty when you put it on the road, and would blow away in a dry period. In some places the good sammel set like concrete – there was a good one opposite Wall End old barn on the Blea Tarn road.

Up on the way to Stickle Tarn, following what we called the green track there is the Little Intake, quite a large enclosure which spans the beck – it just has one entrance from the top and deliberately enclosed the beck for a water supply. The Youdells kept horses and geese in there. It was almost certainly used for collecting sheep before bringing them down for lambing or when transferring the ewes and lambs from the valley to the fells. The ewes would sometimes suffer 'drop' when you put them on poorer grass – it was a help to keep them there until they were better adjusted to the fells again – now we inject them with calcium. Near the Little Intake is a peat house – you would need to dry the peat out up there – there is nothing worse than bringing down a load of wet peat on a sledge with a horse! Between the peat house and the Little Intake are the remains of an old wall. When you are gathering sheep they can come down ten abreast until you reach it when it channels them to only four and slows you up.

Nearby, up on the fells to the east are three small enclosures, like beehives, corbelled inwards, for trapping foxes. You would throw a dead lamb inside. A fox would jump in and would be unable to get out – they have not been used in my time; one is very well preserved, the other two are collapsing. There are many little pens dotted about all over the place – if some of the ewes were slow when you were gathering them, you would put them in there for the night and you knew where to find them the next day, or if a lamb needed extra milk, you would put it in there with the ewe. Above White Ghyll there is a wall to stop the cattle coming down – there have not been cattle up there since we came.

To get up onto the fells from Millbeck we would use the outgang above the farm. Sometimes we would turn right above the fell wall to take the sheep up onto White Ghyll, or up the pony track to the left of Dungeon Ghyll, or the route up by Pike Howe; this was only a sheep track when we came here at first. The new barn at Low Millbeck was built in 1955 from stones off the walls at the back of the farm. The washfold for Millbeck was just above High Tommy, above the present wooden bridge over Stickle Ghyll – we haven't used it for washing our sheep, but in the old days they would have several men in the ghyll ducking them. They would poss 'em about in the water and the sheep would get out themselves! The washfold for Side House was on the old track, the higher track, between there and Oak Howe where the beck meets at the corner of the wall, about a quarter of a mile from Side House.

In a bad winter when there was heavy snow I would take a horse pulling a holly bush up to Stickle Tarn to flatten the snow to make a track for the sheep to follow you down. 1939-40 was a terrible winter – it caused my father to have a breakdown. We were picking up dead lambs everywhere – the ewes couldn't feed. We were feeding lambs all day but they were dehydrated and no good – they died. We were bagging up dead lambs all the time – it was very depressing – and there were no subsidies then. In 1939 they had to cut a way by hand from Chapel Stile to Stool End – it was like walking down a tunnel six feet high – higher in places. At the end of the road, in the tunnel, one day I found a hare – I took it home for the pot!

Lambs were taken from their mothers in the last week in September or the first week in October. Sheep sales started after this – the Ambleside Sheep Fair was on the nearest Fridays to October 13th and 29th. At this time we would drive the hoggs to over-winter from Low Millbeck via Skelwith getting as far as Hawkshead or Sawrey by night time and stay there on the first night and then go on to Grizedale or Haverthwaite or Dale

Park. It was slow going – lambs born on the fells are not the fastest things on four legs! You could reach Dale Park in a day but it would be a real hard day. Other shepherds would come from Langdale, Eskdale, and Seathwaite on the Duddon. The shearlings we would take to Holm Fell and Brow Fell for a couple of months. By the end of the winter there would be 2-3,000 sheep between Hawkshead Hill and Force Forge all ready and wanting to come back – but all needed to be sorted! Mitchells had some land near the Drunken Duck where we kept them overnight – then next day we would sort out the Eskdale and Seathwaite sheep and they would go off together, then the Little Langdale ones would go off with the Fell Foot flock, and then the Great Langdale sheep – Baysbrown, Millbeck, Wall End and Stool End would set off together, sorting out each flock as you reached the right farm.

The hoggs would come home on the first fine day after April 5th. On the way back they were so fit and well fed they could go over anything! The Eskdale flock were taken via Fell Foot where they stayed the night, or you might get as far as Black Hall (Cockley Beck), and then on to Brotherikeld in Eskdale, their own farm, where they were cleaned up, dipped and marked. Occasionally you could get right through to Eskdale in one day. Next day some were back at Fell Foot – they wanted to be back on the good grass – they didn't think much of being back on the fells. All this has stopped now – the Forestry Commission bought the land and planted it, also the people over there started keeping sheep when they found there was money in it. Now the sheep are taken to Lincolnshire to overwinter at £12 per head, it used to be 2s.6d. For the sales we would walk the ewes and wethers to either Ambleside or Broughton – Broughton was a two day trip staying in Coniston or Torver overnight.

At Low Millbeck we start with 320 ewes, plus 40 gimmer shearlings and 40 gimmer hoggs. If you were vacating the farm you would need to leave extra in the autumn before in case the winter was bad, you would need an extra 30 ewes and 10 each of the gimmer shearlings and gimmer hoggs. These were classed as surplus and could be bought by the next man, valued by the viewer, or someone else.

Sheep marks: cropped both ears, red pop on near lisk and another on head.

High Millbeck, since1980 known as Stickle Cottage

This old farm also takes its name from Mill Beck or Mill Ghyll as it is called now, 'High' because it is higher up the valley. The earliest reference known to me is found in the jury verdicts of the manor of Langdale, 1666. John Grigg, tenant of this farm at Millbeck, was one of 49 tenants in Langdale at that time. He died in 1673 and was followed by his son Christopher (d. 1728). The farm stayed in the Grigg family's ownership through John Grigg (1688-1732), his son Christopher (1724-90), who in 1774 had to pay 4s.3d. for land tax, and to a widow Grigg in 1817. In 1843 it was owned by Thomas Newby and occupied by Thomas Middleborough, then George Thompson 1851, George Mackereth 1863-84. In 1885 Joseph Youdell (Dot Parker's grandfather), owned by J.F. Green, and the farm run jointly with the Dungeon Ghyll New Hotel; Joe Cowperthwaite 1907, John Fothergill 1919. Guy Senior bought the farm from Mr Fothergill 1925, then Cyril G. Bulman 1941, W. Miles Jessop 1973, and David and Mary Quine 1985.

The whole estate including the hotel and Rossett was up for sale in 1907; the highest offer of £2,000 was turned down and the sale withdrawn.

The house has many old features – the huge oak beams, some with pieces cut out for jointing showing that they were re-used from an older building. Some of these beams have a natural curve which gives the upper floors an undulating surface. Some of the oak floorboards, with wooden pegs are still in place. To the right of the main fireplace at the east end of the house is an old spice cupboard with butterfly hinges and a primitive rough catch, in the corner there was probably a fire window which was blocked up to avoid the window tax – a cupboard in the wall still exists, with a shelf over it to keep the family Bible. In the loft there is a wooden lintel, one foot above the floor – this was probably over the doorway into the back barn of the original house.

Round the back of the house huge slate slabs were propped up against the building and covered with stones and earth providing a very efficient air vent before damp-courses were invented. A map in 1852 shows a garden in front of the house and one on the east side with a large orchard behind. The orchard has been replanted in part in 1989. Attached to the house to the west is a lean-to milking parlour, still with the remains of the stalls and the hooks for tying up the cows. At the back of the house another lean-to provided a wash-house and a fuel store.

The farmhouse was also an ale house, before the Dungeon Ghyll New Hotel was built in 1862, and had four bedrooms upstairs and an outside toilet on stilts reached from upstairs. The present Stickle Barn, now a restaurant, was the barn for the farm, and the building next door was the stable block with barn above before it was converted into a house in 1955. Behind this, the present bunk house, sometimes called the 'Long Tram' was an open coach house with pillars along the east side, used for keeping the horse traps and coaches when the farm functioned in conjunction with the hotel. In the summer three coachloads of visitors would stop for tea and a walk up to the Dungeon Ghyll waterfall, while the horses had their food and a well-earned rest. To the west of the coach house were the pigsties and a small pig field. It had been a flourishing farm in its time.

The fields attached to the farm were bounded to the east by Mill Beck (except for Old Close in 1852), Langdale Beck to the south and reaching to Rossett Bridge to the west. The names are Back Meadow, Great Meadow, Great Croft, Little Croft, Appletree Howe and Mill Close, Low Tommy Close, Low Crag and Stones, the Howe and Beckside (1852 Rossett Backing), Stony Lands, Low Close, Thorns Close, Grubbin, Kirk Howe and Tommy Close (1852 Tovey Close), New Intack, and Fold Howe Intack (with the remains of the larch plantation).

Dungeon Ghyll New Hotel

This is a comparatively new hotel, a datestone in the pavings in the front entrance gives 1862. The original inn was in the old farmhouse at High Millbeck. Occupants of the hotel were George Mackereth who died there in 1884, followed by Arther Knubley in 1885. Joseph Youdell took over until he left in the spring of 1908 and was followed by Joe Cowperthwaite and then by John Fothergill. When he was at the hotel he had a horseman called James Bainbridge, who always went by the name of Jimmy 'Ledder' (possibly because he was as tough as leather!). He was a bad-tempered old fellow and when he was driving the horse and coach up the valley we used to run behind to get home quicker, but he used to whip you off if he realised you were there! Albert Black was there until he sold it in 1949 to Cyril G. Bulman. Miles Jessop bought it and made the farm buildings into a cafe and bar. Thc hotel was then sold to Mr Booth who leased it to Quest Adventure Holidays (although the bar was independently run), and the restaurant sold to Graham Hill and his partner Mr Jinks. In 1989 Graham and Kathleen Hill became sole proprietors of the Stickle Barn restaurant and in 1992 sold it to Peter and Sylvia Ingham. In 1991 the brothers John and David Smith bought the hotel and have worked hard to restore it into an attractive country house hotel.

One hundred years ago, the hotel was described in an old guide book for 1891:

> Dungeon Ghyll New Hotel, most conveniently situated for the ascent of Scafell and Bowfell, or for crossing the passes to Wastdale, Borrowdale etc. The best guides, conveyances, and mountain ponies may be had at the Hotel, also for all information respecting mountains and mountain passes. Board and lodging (except August) 35 shillings a week.

Prices have changed a little since then!

In the early 1900s Langdale had its own football club which played in the field in front of the hotel.

Side House

Side House is another farm which stands on the sunless side of the valley, built in the usual style with a piece added on in about 1936. It has a good range of farm buildings which were erected in quite recent times. There is a plaque on the wall in the memory of Professor G.M. Trevelyan acknowledging his love for the Langdale valley.

Occupants were John Grigg 1661, John Dixon 1728, Edward Satterthwaite 1790, Joseph Grigg 1794, Thomas and Margaret Milligan 1871, William Birkett 1898. Thomas Birkett (son of William) was here in 1900, married Ann Chapman and had four children, Mary, Robert, Ruth, and Annie; Robert married Mary Myers, the eldest daughter of Ezekiel

Myers of Wall End, in 1925. After old Tom retired, Robert and Mary carried on the farm until 1940 when they left and went to High Yewdale in Coniston. After they left, Side House was taken over by Joe and Beattie Park who farmed there until 1948 when they moved to The Oaks Farm, Loughrigg. The farm was bought by Kenneth Blackburn who had Ernest Dent as the farm manager until it was sold up in 1963. The farm was given to the National Trust as a memorial to Professor G.M. Trevelyan and it ceased to be a separate holding. The land was put to Millbeck and the sheep went to Middlefell.

At the clippings when the local farmers sang their favourite songs, old Tom Birkett invariably sang 'The Dolly Varden Hat' – I can still hear him singing this song as if I had it on tape. The chorus went:

> Come dear, don't fear, get your ringlets curled;
> If you aren't in the fashion, you'd better leave the world.
> Bright and pretty faces wear a winning smile,
> And wear a pretty hat in the Dolly Varden style.

Sheep marks: under key bitted both ears and a red pop on the far lisk and one on the tail head.

Field names: Corn Close, Crabtree, Far Meadow, Horse Pasture, Low Field, Middle Field, Dungeon Cow Pasture, Top and Bottom Intacks.

We in the Langdales don't get many thick snows – we do on the fells but not in the valleys. It so happened on 25 February 1947, when Joe and Beattie Park were at Side House and were expecting their first child, they awoke to a very deep covering of snow. The roads were impassable and the nurse had to be conveyed by horseback – it would have to happen on this particular day. However, the baby was delivered safely and was named Dorothy. Dorothy now lives at The Oaks in Loughrigg with her husband, Alan Wrathall, and family of three. Strangely, she is now in the house where she spent a good deal of her young days.

Rossett

Rossett is a beautiful old-world house with stone-flagged floors and a lovely stone staircase with arc-shaped steps on the turn and with rocks jutting out from the walls. It has had two pieces added in recent years. It has a very old barn attached to the house at the back with marvellous cruck beams, thought to be built around 1550, with the house about 100 years later. On one early map it is called Rossett Sheiling and was the place where the packhorse ponies rested after their journey over the pass. It gives its name to Rossett Ghyll rather than the other way round. The pass itself was constructed between 1690 and 1720.

This has been a small farm with some fields; no doubt the field in front of the house has belonged to it at some time as the approach road divides it. Some fine old trees, including yews, stand in the grounds and give some protection from the elements.

It has had a stock of sheep and I have an old bond dated 15 April 1735, in which Robert Jenkinson of Hinning House in Seathwaite and Edward Benson of Millbeck in Great Langdale were sureties for John Dixon of Rossett in Great Langdale, concerning the management and redelivery of sheep at Rossett.

In 1684 a Rydal man presented before the justices quite a concourse of people who had been present at a conventicle in Langdale, some 17 Loughrigg and Langdale names were cited including Edward Benson of High Close, John Dixon of Rossett, and William and James Harrison of Harry Place. The Quakers long held meetings at Rossett.

In 1881 William Walker lived here with his wife Mary and their large family, three children to her first husband and four more to William. There have been numerous tenants living here. Guy Senior bought it in 1925 with the stock of sheep and lambs that went with High Millbeck which had latterly been run by Mr and Mrs Fothergill of the New Dungeon Ghyll Hotel. The farm man who lived in the old farm house was Art Askew, and when the sheep were sold he went to farm at the Pike in Seathwaite. Guy Senior farmed Rossett for about a year when he sold the sheep to Harry Mounsey who was farming at Low Millbeck. The house at Rossett was sold to R. Holden in 1926 and is now owned and lived in by his son John and wife Joan.

Dungeon Ghyll Old Hotel

The Old Dungeon Ghyll Hotel was originally a farm and an inn. In the *Shepherd's Guide* of 1879 it was named as Middlefell Place run by J. Bennett and in the rate book of 1885, it was named as Middlefell Inn and run by the same John Bennett, who was a well-known guide for tourists.

It has been extended a number of times, the original piece being the middle portion; another piece was added later and the stables at the north-east end were made into a dining room. The shippon, on the other end, was made into a bar and the cow stalls are still there, intact.

Sheep marks: cropped near ear, upper shear, halved far with two short strokes over fillets and a pop on the shoulder.

The fields were the meadows just over Slops Bridge (Rossett Bridge) and I can remember when this land was not fenced by the roadside. There were also one or two fields near the hotel but all this land has been re-allocated.

John Bennett sold the hotel in 1888 to William Baines who was there for three or four years. Joseph Youdell then took it over, running both the Old and the New Dungeon Ghyll hotels. His wife, Margaret, who looked after the New Hotel, died in 1907. They had a large family of 14 – seven boys and seven girls – Wilson (Wilse), Thomas, 'Faif' William had a humpback, Joe who was lost in the First War, Edward (Ebert), Fred, and Percy; the girls were Annie, Louise (Weety), Mary (Molly) the mother of Johnny Holmes, Hannah, Margaret who became Mrs Wood, Emma who remained unmarried, and Bessie who became Mrs Thomas. This large family made it possible to run the two hotels. These were the days of the horse-drawn 'charas' bringing visitors from Little Langdale over Blea Tarn Pass; they would stop at the top and blow their horn, a signal to get lunch or tea ready – the number of blasts informed the staff of the number of passengers requiring the meal! For many years, when Thomas Henry Youdell (Johnny Holmes's uncle) was at the hotel, he used to go to the top of Esk Hause to collect the rainfall from a copper rain gauge on the last day of each month. The water had to be taken to James Marshall at Skelwith Fold. He was one of the family who owned Monk Coniston and Patterdale Hall, and he built the tarn by the roadside to Coniston; as a side line he was interested in the rainfall and kept records from Esk Hause and probably other places as well. Remuneration to Thomas Youdell was one golden guinea or thereabouts on a once-a-year basis.

Anthony Chapman, of hunting fame, came and he left in 1915, but not before winning the reputation for unseating three Members of Parliament in one day. Perhaps a little worse for drink, he was taking the three back to Windermere to catch the train, however, at Slops Corner he managed to turn the trap over and unseated the three gentlemen all at once. Joseph Holmes (Johnny Holmes's grandfather) had the hotel for a time before selling it for £4,100 to Professor G.M. Trevelyan who promptly gave it to the National Trust. This was the first property owned by the Trust in Langdale. Mr Airey occupied the property but he did not stay long – I can well remember getting rides up from school in the waggonette drawn by a white horse and driven by Mr Airey's brother. Next came John Dawson, from 1922-37; he was responsible for making the present dining room out of the stables. Cyril G. Bulman followed in 1937 until 1949; he was the last to run the hotel with the farm; he then went to the New Dungeon Ghyll Hotel which he had bought.

Sid Cross, a well-known and experienced rock-climber, came from Eskdale with his wife Jammie in 1949 and stayed for 21 years. Sid was already known in Langdale, but not being a farmer he didn't require the land. They moved in on a rather wet, miserable, uninviting morning; before long, there was a knock on the back door, Joe Gregg was there, 'I've come to wish you good luck. Glad you've come to the valley. What are you going to do about milk? I could supply you'. Sid agreed. Soon after there was another knock at the door. This time it was Jim Mitchell. 'I've come to wish you good luck. Glad you have come to the valley. What are you going to do about milk?' Sid commented 'It was a wonderful welcome we were given by everyone in the valley'. Not being a farmer Sid employed William Bowness – Uncle Will – to do all sorts of jobs around the hotel, he was always willing to look after the hens, pigs and even the youngsters! One of the first things they did was to convert the shippons into the Climbers' Bar, retaining as many old features as possible, it was ready for their first Easter, together with toilets in a nearby building.

With the beck nearby Sid was to find flooding quite a problem; often the cellars were under 2 feet of water after heavy rains.

One Christmas, the first one the hotel was open for residents, Joe Gregg came and said 'You'll be needing extra milk, cream and butter won't you? I'm going to the auction and I'll get an extra cow'. Sid said 'You can't get an extra cow just for us!' Joe replied 'You run your business and I'll run mine!'

All the climbing clubs in the country at this stage came to the Old Hotel for their club dinners which meant that many of the best British climbers who had taken part in the Everest and other Himalayan expeditions stayed at the hotel including Sir John Hunt, Mike Westmacott, Alf Gregory, Tom Bourdillon, Sir Charles Evans, Neil Mather, and George Band. The younger up-and-coming climbers frequented the Climbers' Bar – Joe Brown, Don Whillans, Ian MacNaught-Davis, and Dennis Davis. In the evenings they often had talks and slide shows – Chris Bonington and Ian Clough, after being the first British climbers to ascend the north face of the Eiger, tried out their lecture at the Old Hotel before giving the real thing in Keswick.

One morning at 8am Jammie and Sid received a phone call from Tokyo, it was just before the Emperor died, and the caller mentioned that the Emperor of Japan had stayed at the Old Hotel when he was the Prince and he was wanting to know the year. After some research Sid was able to tell him that it was on 3 September 1915 that he stayed there and climbed in the area. These were great days at the Old Hotel and Sid and Jammie were sadly missed when they left. Mr and Mrs Carr took over, now Neil and Jane Walmsley are developing this popular family hotel – Neil is a talented musician and a keen fell runner.

Middlefell Place

Middlefell Place is situated at the head of Great Langdale where the road starts to go up Mickleden, and must get its name from its position in the middle of the fells. There were a lot of people called Middlefell too, including Charles Middlefell who was reader at Langdale old chapel – he died in 1643. The farm is in a good position apart from being on rather low ground and a bit too close to the beck which has been known to overflow and get into the house on quite a few occasions.

The house is an old-world one and has been sketched and photographed many times in its setting with the sycamore tree and Bowfell standing majestically behind. There are no yew trees round the farm. The house has five bedrooms but no particular features inside, no spice cupboard or court cupboard.

There were two or three holdings at Middlefell including the present Old Dungeon Ghyll Hotel, once called Middlefell Inn. In an old rate book for 1885 it refers to Middlefell Inn being occupied by John Bennett and owned by Thomas Blamire. Another entry in the same book refers to another 'house and land'. The third entry, which is for Middlefell Place Farm, states that it was occupied by Noble Bland and owned by Jacob Gillbanks.

The old hump-backed bridge – a packhorse bridge – which went across from the main road is still in existence; close inspection shows that it has been widened at some stage. The new bridge was built in 1925.

There are also the remains of a farm called Ash Busk, down by the river where the ash trees flourish. It is approached from an old walled lane off the Mickleden road, two fields north-west of Middlefell. From records a Robert Harrison lived at Ash Busk as early as 1690 but the farm has long been a ruin.

The Harrison family were very strongly represented in Langdale in the early days and Harrison Stickle was no doubt named after them along with Harry Place and Harrison Place, now called Robin Ghyll. There were Harrisons at Stool End, Raw Head, Pye Howe, Ellers, and Blea Tarn; in Little Langdale at The Busk, Dale End, The Brow, Thrang, and in Loughrigg at The Oaks, The Howe, The Fold, The Ellers, High Close, and Tarn Foot.

Occupants of Middlefell Place were William Dixon 1663, John Wilson 1683, Charles Middlefell 1698, John Otway 1701, Charles Middlefell 1707, John Gillbanks 1767, Joseph Postlethwaite 1817, Noble Bland 1885, Jerry Grisedale, followed by his widow, and then their son Herbert, John Youdell 1943, Bryan Tuer 1953, Hugh Parker 1961, Mike Toms 1976

When old Jerry Grisedale died at Middlefell in 1919 I can well remember seeing the wheeled bier going down the old road with the corpse. It was a good three miles each way that the undertakers had to walk. This was the first service at which my sister Hannah played the organ.

There used to be 360 sheep and the marks were upper key bitted and under halved near ear and a black pop on fillets; ewes a black pop behind the shoulders.

Field names: Ling Ley, Mucked Earth, Broad Piece, Crook Holme, Hag, Low Above House, High Above House, Ash Busk, Side Intack, Beck Stones, Blake Howe, and Willy Door.

Stool End

This holding stands at the head of the valley at the foot of the Band and has variously been named Stool End, Steale End, or Style End. The house is built in the traditional way and has four bedrooms and quite a good range of farm buildings, some old and some new, together with a stock of Herdwick sheep which go on the Crinkle Crag range. The river which passes nearby comes down Oxendale and in rainy times can be rough as it has a good gather from the surrounding fells. In 1966 the Stool End bridge was washed away in the floods.

There used to be four farms at Stool End. There is mention in an old document of an old homestead called 'Stangs' which might refer to tenters or poles upon which they dried their cloth – where it was stretched and dried. In this case there might have been a walk mill not far away.

The farm came up for auction at the Salutation Inn in Ambleside on 12 July 1792. It was described in the following terms:

> A beautiful and very improveable Estate called
>
> STYLE-END
>
> Situate in Langdale
>
> Which abounds with pleasant, extensive, and entertaining Prospects,
>
> Within a few Miles of Ambleside and Hawkshead, two good Market Towns. Not far from the much admired Lakes of Connistone, Windermere Grassmere and Ulswater:
>
> Consisting of Three Dwelling Houses, three Barns, and other suitable Outhouses, all slated and in good repair; and containing (by Estimation) two hundred acres, or thereabouts, customary Measure, of arable, meadow, and pasture Land, well fenced and in good Husbandry, all laying together, and adjoining to an excellent Common, on which it has an unlimited Right, and on which a Thousand sheep may annually be kept; holden by Payment of several small yearly customary Rents, as part of the Marquis of Richmond-Fee, and pays only three Times those Rents for a Fine, and on Change of Tenant only.
>
> William Holme, the present Farmer, will shew the Premises and for further Particulars apply to Messrs. Thomas Garnett, Robert Hutchinson, Christopher Longmire, and John Nicholson, Devisees in Trust under the Will of John Longmire, Maltster, deceased, the late Owner; or to Mr. Postlethwaite, Attorney, in Kendal.
>
> N.B. Four Hundred and Eighty Heath-bred Sheep are now let along with the Estate which the Purchaser may have at fair Valuation.

In later years the stock with the farm numbered 358. The sheep marks were square forked near ear, under key bitted far but in an older *Shepherd's Guide* it was cropped and muck forked near ear, under key bitted far, with a red pop on the back.

Field names: Hackside Close, Little Intack and Thin Ing, Little Intake, Ash Busk, Sheep Wash Intake, Far Lea Close, Near Lea Close, Lob Intack, Little Moss Dyke, Great Moss Dyke, Bull Copy, Far Goody Close, Near Goody Close, Barn Close, Low Lands, Hoghouse Intacks, Great Crag Intack, Little Moss Intack, Little Moss, High Lands, Boon House Closes,

Green, Rough Intacks, High Intack, Oxendale Low Close Intacks), Drum, Crook Holm, Low Wood, High Wood, Broad Dales and Flood Leas, Parrock, and Applegarth.

Occupants were Reginald Wilson 1625, John Harrison 1673, George Benson 1707, John Grigg 1710, John Longmire, maltster, had owned it, now deceased 1792, William Holme 1792, Mark Wilson 1794, Mr Sandford 1817, John Grisedale 1871. Thomas Martindale, who bought the farm in 1876, died here in 1889 and was followed by his son William. He married Ann Crossfield and had a daughter, Margaret. From 1919 Jim Mitchell, he had lived with Billy Martindale from the age of 14 years and took over when Billy retired. Jim married Dora Allonby that year and had a daughter, Elizabeth, born in 1920 and a son, John, born in 1922. He farmed here until he retired in 1963 when Harry Cowman came from The Busk in Little Langdale. He left in 1973 and was followed by Keith Rowand.

Wall End

This farm stands at the foot of the Blea Tarn Pass at the head of Great Langdale. It is built in the usual style with a good range of farm buildings which are attached to the house. It is quite a large dwelling with five bedrooms, and a piece has been added later for a kitchen. There is a short time during the dead of winter, from the end of November until 15 January, when the sun does not reach the buildings. Three large yew trees stand in front of the house and are much too close to it making the rooms very dark.

Inside the house are some lovely oak beams and panelling with a beautiful court cupboard inscribed RGIG 1725 (John Grigg and his wife). This is situated on the right-hand side as you enter the front door – it stands in the passage with its back in the right-hand room. On the left is the sitting room with a good spice cupboard with the initials G over E.S. (Ephraim and Sarah Grigg). In this room the windows have wooden shutters, which is unusual. Approximately 5 feet above the floor and cut into the dividing wall between the two main windows are two small horizontal tunnels, about 5 x 5 inches in section and nearly 2 feet long. These lead in a curve to two little window panes, each about 6 x 3½ inches, in the outside front wall of the house. They are hidden from view on the inside when the shutters are open but are clearly seen on the front of the house. Perhaps it is a double peep-hole window; it gives a good view of anyone approaching either of the front doors. It is certainly unique in the area. When Mary Birkett lived here as a little girl she found it very useful for hiding her sweets! Jonathan Myers informed me that it had been covered in at some stage but was rediscovered in 1950 when the walls were picked off and re-rendered

Belonging to the farm is the old barn at the foot of the Blea Tarn Pass. It has recently been re-roofed in the traditional way with riven timbers and wooden pegs. It has a marvellous set of three pairs of cruck beams, some a foot square in cross-section; it also has side purlins and the ridge purlin is carried on a king block and saddle.

Kettle Ghyll comes down past the house and can be very rough after heavy rain. Soon after Jim and Greta Edmondson came in about 1966 there was a very severe storm which built up the beck with boulders and diverted the river down the road and into the house. It took days to clean it up. This was the same flood which washed the bridge away at Stool End and did a tremendous amount of damage down the valley.

In 1929 the farm was bought from John Caddy Nicholson by Professor G.M.Trevelyan who gave it to the National Trust.

The farm has a stock of heaf-going sheep which used to be 457, with these marks – ritted near ear, cropped far, stroke over the fillets and a pop on the far shoulder. In one *Shepherd's Guide* it is described as the Oxendale stock. Until 1966 a number of two-shear wethers were kept in the flock to maintain the heaf, this was disbanded because it was no longer viable and was substituted by ewes.

Field names: Hogg Platts, Broad Dale, New Meadow, Back Field, Garth, Long Dale, Long Meadow, Top Holmes, Low Holmes, Bull Field, Little Meadow, Sides Intack, Broad Intack, Long Intack, and Wash Dub Field.

In an old document dated 1625 'Raignold Wilson, sonne of Richard Wilson of the Waleend in Langdal, husbandman [sold] an intacke called the Peathow in Langdall near to the Crostake' for £5. Crostake, may refer to Crossdyke, the manorial boundary of Baysbrown where it crosses the present road to the north of Blea Tarn.

Ezekiel Myers, from Torver, was in mining, quarrying, and farming when he married Jessie Richardson from Borrowdale. He left his wife and two daughters in England while he went to look for work in America. However, when he returned for his family Jessie declined to go abroad with him and they went to Wall End instead. They arrived in 1913 with Mary, who married Robert Birkett, Jessie who married Llewelyn Jones, and Isaac. The rest of the family were born at Wall End; Margaret, Sarah, Helen, and Ezekiel ('Zeke' Junior, the inventor!).

Isaac took over Wall End in 1946. He married Hannah Greenhow of Low Colwith and their family consisted of Jonathan who married Sheila Taylforth, Isaac who married Ruth Clarke of Oldham, Ezekiel, a joiner, who married Judith Oliphant of Coniston, Helen who married Brian Jackson, the farmer at Oxen Park, and Ann who married Tony Sanders, a blacksmith at Torver. Isaac, the father, and his brother 'Zeke' were very inventive, after the Second World War they used ex-army surplus equipment in an attempt to develop a hydro-electric scheme up Kettle

Ghyll. It was only partially successful, but they did manage to clip sheep by the means of water-power. The scheme was disbanded when they obtained a stationary engine-generator unit. They also made a motorised mowing machine with iron wheels, which was used for several years, out of an old Standard car and a horse-drawn mower. Zeke was well known as a motor cycle scrambler and, with an ex-works Norton, was for several years one of the top men at this sport in northern England. He also started a haulage business by taking cows to market for the farmers and this developed well. He married from Wall End and moved out of the caravan he had been living in to take up residence at High Close.

Blea Tarn House

Wordsworth called it the home of the 'Solitary' – 'one bare dwelling, one abode, no more'. This is an old farm holding with access from Great Langdale over Blea Tarn Pass or from Little Langdale up Horse Brow. It used indeed to be a very lonely abode, but now it is seen by hundreds of walkers and people in cars. Years ago the old horse-drawn 'charas' used to go up Little Langdale and down Great Langdale but the passengers had to get out and walk up the steep hills. They had their own time-table and usually called at one of the Great Langdale hotels for tea.

The farmhouse is built on the usual pattern and you can see the tarn from one of the bedrooms – you get a better view when the leaves are off the trees. There are no yew trees, but there is a large leaning larch at the end of the house and mature trees nearby. Tawny owls are quite often seen and heard in the area. The farm has been photographed and painted many times with Bowfell in the background. Lonely, yes, but you would never call it 'nebber thrang', which means that no neighbours ever popped in. It is a good place to call for walkers who are thirsty.

There is a wonderful set of stone stairs which is quite unusual as it goes all the way to the landing. There are some lovely oak beams inside which are well preserved – some on the landing are huge. There are five bedrooms, originally there were only three, but a part of the house has been added on. The chimneys are in the usual squat pattern and the old porch is in the Lakeland style. It stands in quite a sheltered position and is right on the roadside. In the barn there are some very old half-cruck beams.

Blea Tarn is no longer run as a farm and the land was added to Fell Foot Farm about 1972 after the farm was given to the National Trust.

Field names: Tip Close, Paddock, House Field, Moss Field, Gale Meadow, High Intake, Low Intake, Ling Moor, Raw Crag.

Sheep marks: cropped near ear and under halved far; smit marks – a pop on near lisk and one at tail head.

Occupants were George and Mayson Weir 1885, Thomas Barrow 1899,

Mayson Weir in First World War, then Tom Taylforth and his brother Bob who went to farm at Elterwater, Alfred Webb 1933 (his sister Nellie had the toffee shop in Chapel Stile), then Tim Redhead and Bob Taylor, Phillipsons 1947, Noel Hodgson 1958, Keith Rowand 1972, Mike Toms 1974, Jonathan and Sheila Myers 1976.

Where you have communities of people you invariably get the aggressive or quarrelsome individuals. This happens today and certainly happened long ago. About 75 years ago Mayson Weir farmed Blea Tarn – he had rather an aggressive nature. His wife was sister to Jackson Youdell's wife who farmed Fell Foot, which is not a great distance from Blea Tarn. After some sort of a quarrel Mrs Youdell had thrown a bucket of water at Mayson. Some time later Mayson, armed with a bucket of water, left Blea Tarn en route for Fell Foot to get his own back. He went and knocked at the door and when Mrs Youdell opened the door he threw the bucket of water right inside the door and down the passage saying 'There's your bucket of water back!'

Wall End

CHAPTER 12

FARMS IN CHAPEL STILE, ELTERWATER AND COLWITH

Walthwaite Farm (1), now Walthwaite How

This was a small farm consisting of a lovely old house with a separate barn and shippon attached and a cottage on the north end. The farm has been altered a great deal with back buildings added to the house which had a beautiful court cupboard with carving WSA 1711. This has recently been removed. Legend tells of a murder taking place in this farmhouse many years ago. The book *The Bondswoman* by W.G. Collingwood is based on this old farm – he also mentions Thrang, Rossett, and Baysbrown.

The house and buildings stand next to the road and enjoy beautiful views across to Wetherlam and Lingmoor and the surrounding countryside. My forebears, the Stables family, were brought up here. My grandfather John Stables and his father Henry before him had their shoemaking business and also ran the farm. It had a small stock of sheep to run on the common and, according to a *Shepherd's Guide* of about 1800, Matthew Spedding registered sheep and the mark was under halved near ear, under key bitted far, with a red half crupping on the near side. It was run as a small farm as recently as 1959, William Wood being the last tenant. It then belonged to Mary Alice Stables, and before that to her father James. My brother John Fletcher Buntin bought it off her in 1959. The house needed a lot of repairs so he sold it to Clifford Fish who altered and renovated it, while my brother kept the land, but he died in 1960 and I now own the land.

The land belonging to this holding was no more than about 14 acres and, as there have been a few houses built on some of it, it is now even less. The parrock at the back was sold off with the farmhouse and two new houses have recently been built and the small barn at the back of the house has been created into another dwelling.

Walthwaite Farm (2), now Walthwaite House

This is a beautiful little pink-washed farm on the south side of the road a little further out of Chapel Stile. It stands at the top of the land which belongs to it with a parrock at the top side of the road. The house and outbuildings are in a well-secluded and sheltered position with the usual yew trees which give a good deal of protection. A small cottage is built on the south end.

The land is in a ring mostly fenced by walls and extending to about 20 acres plus the parrock of four acres. The names of the fields are Hoggus Field, Culpit How, Nickle Meadow with Round Nork, and Bottom Pasture. This small farm had a stock of sheep to run on the common. The mark in the *Shepherd's Guide* of 1800 was ritted near ear, cropped far, and a red crupping on the tail head, and a pop on the near shoulder. This was entered by William Spedding, no doubt a relative of Matthew Spedding who ran the other farm next door. This farm was owned by Thomas Thompson in the 1800s and was run by Jane Ann Thompson. It is now owned by her relatives, the Pearce family, the house being kept as a holiday house with the land let off. The land at the south-east called Cylinders, about 12 acres, was owned by the Elterwater Gunpowder Company and is now owned by Harry Pearce.

Walthwaite was quite a community of houses in years gone by and I have heard from an old resident that about 30 children went to school from here. There were some humorous people who lived in the community. The village tailor was one of these – Benson Tailor. He was somewhat of a character. One day when he was walking near the slate quarry he met one of the quarry workers. As it happened, Benson had a button off his jacket and the quarryman noticed this and commented 'It isn't often you see a tailor with a button off'. Benson was quick on the mark and retorted 'No, it isn't often you see a quarryman with a slate off!'

The Thrang

At the Thrang there were two little old farmsteads, situated on the edge of Chapel Stile. The road which turns up near the Grassings and goes behind the present school went past them and then carried on down a very narrow old lane, over the new bridge and on to Baysbrown. Both the old farms had other buildings attached as was the custom. The two farmhouses have been modernised and improved in later years, and one of the barns has been made into a dwelling.

The land attached to these holdings was quite small, no more than about 16 acres. However, in days gone by these two holdings would be occupied by two families and no doubt they managed to make a living there. I cannot find any record of a stock of sheep at either of these places. They are certainly old – there is a record of a William Gollinge 'of the Thrange in Langden, clerk' who was a reader at Langdale church in 1590.

Isaac Bird was the owner of one of these holdings latterly and no doubt Bird Hill as we know it was named after him. Later it was owned by Isaac

Bird Gaskell, known as 'Dicky Bird' – he died in 1922 and the farm was later left to Miss Rachel Gaskell who taught in Great Langdale school and was my first teacher. She was born in 1886 and died in the early 1930s.

The Grassings, now Wainwrights

This was a small farmstead with a barn attached until it was made into the Langdales Hotel by William Smith in the early 1930s. The licence and business was transferred from the White Lion which was next to the church. It has been altered out of all recognition. The old house had a barn attached to the end. The farm and 12 acres of land belonged to the Elterwater Gunpowder Company, and in the *Shepherd's Guide* two stocks of sheep are mentioned, the marks being, for the Low Fell stock – forked near ear and black pop on the top of the shoulders, and for the High Fell stock – forked near ear and under key bitted both and a black pop on shoulders with red stroke over back behind shoulders. There was a public washfold just near the premises by the bridge. I can still picture this old building – I find it sad to see these old buildings desecrated like this. This is sure to happen and I suppose one should not hamper these developments if they are good for the community. The Grassings is now the Wainwrights, a pub, together with self-catering accommodation – part of the timeshare complex.

Old Hall Farm, Kitty Hall, and Lane Ends, Elterwater

Old Hall Farm was mainly a dairy farm – there are no records of sheep marks attributed to it. It stands on the narrow road to Colwith as it leaves Elterwater. It ceased to be a farm in about 1937 and the farm buildings were made into a Youth Hostel about that year. The farmhouse with four bedrooms is now a cottage; it used to have a date stone on the front – 1692. The land, with the exception of one or two fields, runs down to the south-west of the river and to Elterwater Tarn.

In 1885 the farm was let to Thomas Grindley and owned by Mrs John Robinson. John Dixon died here in 1895. William Parker farmed here in 1918 and he was followed by Harry Barrow and Robert Taylforth until it ceased to be a farm. In 1936 when the Elterwater estate was sold up it was bought by Mr Pattinson for £1,050.

Kitty Hall has no doubt been a small farm long ago and would have had a few small fields. It has a barn more or less attached to the house. It is very ancient and of rugged construction; the person presently living there says it is perfectly dry inside. It has three large bedrooms and stands in a very secluded position. The land extends to about 80 acres and includes the tarn. The land over the other side of the river belonged to the Gunpowder Company and a path leading from Elterwater goes to Skelwith Bridge – a lovely walk by the river and the tarn and through the beautiful woods, carpeted with bluebells in the spring.

Lane Ends was another very small farm with the land adjoining these fields. The barn which accommodated this farm has been made into a cottage and the National Trust owns the land along with the Gunpowder Meadows.

High and Low Colwith

These two farms are situated at the junction where the road from Elterwater has the turn off to Little Langdale. High Colwith, on the top side of the road, is an attractive little farm with barn attached, right by the roadside. It has a kitchen, living room, and three bedrooms. Some of the land joins the Hacket land which now goes with High Colwith. In 1885 it was farmed by William Bates, a cattle dealer and guide to Colwith Force. He held the key and for a small payment you were allowed to visit the waterfalls – the path is now entirely free and the falls well worth a visit. Later the farm was run by Mr and Mrs Abbott and until recently by Mr and Mrs Ron Young; it is good to know that it is now let to a local couple, Mr and Mrs Neil Hodgson.

Low Colwith is on the opposite side of the road, on the left as you come from Elterwater, an old house with barn attached in a very secluded position with some of the land running down to the river. In 1885 it was farmed by John Dixon and later by Jonathan Greenhow. I can still picture his wife, Maggie, wending her way to Ambleside in her horse and trap with some of their own produce – butter, eggs, and cut flowers. Times were hard and you had to make a few coppers when you could. Their daughter Hannah married Isaac Myers and they ran Wall End for a number of years. Now this small holding is lived in by Philip and Susan Bowness, Philip running his business, 'Colwith Interiors', from the barn. His twin brother Richard with his wife Hazel live in the modern cottage opposite, from which Richard runs his building business. In 1849 one of these farms was worked by Lanty Slee, the whisky distiller. There is no record of flocks of sheep at these two farms.

CHAPTER 13

LITTLE LANGDALE FARMS

High and Low Hackett, and Iving Howe

The two farms at Hacket are approached off the Langdale to Colwith road. Turn up right at the top of the hill past High Colwith, as you go up a winding old road firstly you see Low Hacket. It is rather a large house which has been rebuilt and extended. Whether any of the original structure is left I cannot say. It is now owned by Dr Gregson. My grandfather had this land 'taken in' and my mother used to tell me how more than a hundred years ago in 1888 they used to go haymaking there. At one time Low Hacket was owned by Langdale church with half of High Colwith and most of Bridge End Farm. There are many conveyances held at the church for these properties.

High Hacket is a little higher up the valley and is approached by an old winding grassy track. The views from these two farms are magnificent. The land is full of craggy knolls and is very undulating, but also sheltered by lots of trees. In the 1880s this farm was owned by Reginald G. Wordsworth and run by Michael Rigg.

Iving Howe or Ivy Howe as it is often called, was also owned and worked by the same two men. This was a small farm situated on the opposite side of the road to the forge; it has been renovated and extended many times. It is said that Lanty Slee built a small house at Ivy Howe from the proceeds of selling his illicit whisky. He seems to have lived and operated from various houses in the area! They say 'a rolling stone gathers no moss' but Lanty seemed to make a 'canny bit o' brass' even 'though it was only 10 shillings a gallon'. It is also said that he bought a small farm called Greenbanks from the profits from his whisky, where he died in 1878 at the age of 76.

Hacket Forge

On the roadside from Colwith to Little Langdale stands this ancient farm holding dating back to 1608. This was the place where stood an old bloomery – an iron-smelting forge. It was built right on the edge of the river and traces of charcoal can still be dug up in the garden. The iron ore was transported here from Red Tarn near the top of Wrynose (some 6,000 tons of ore were extracted), and also from the top of Rossett Ghyll in Great Langdale. In fact the Rossett Ghyll track was developed between 1690 and 1720 by the forge-owners to bring down the ore. Some 30 tons of iron were produced in good years.

The earliest reference is of a sale by Joseph and William Pennington to William Wright of Cunsey in 1623 for £250 which included the surrounding trees. A few years later two fulling mills were converted into a forge by William Wright now of Easthead, who had also purchased from John Wilson of Little Langdale the tenantright of a piece of ground called 'Ivinhoe within the Hacket'. The river at this time was called Haccat Water or Langdale Water. By 1652 the forge was worked by Gawen Braithwaite who bequeathed it to his son Thomas, both working on behalf of Sir William Pennington. By 1710-11 the forge was worked by William Rawlinson and John Machell who produced 19 and 11 tons in those years. It was no longer a bloomery forge but used for refining pig iron into bars. The pig iron was brought up Coniston Lake from Backbarrow in a boat from Nibthwaite to Waterhead at 1s.3d. per load and then by packhorse to Langdale for a further 1s. per load. Work there seems to have stopped around 1720.

The house has some lovely oak beams and a marvellous stone staircase up to the bedrooms. In the past, the kitchen was used for village dances, complete with the fiddler. The barn is adjacent to the house and the farm was run alongside the smelting works. It is very ancient. The presence of pitsteads where the charcoal was produced can still be found at Hacket Forge indicating that the bloomery was self-sufficient. In 1565 all the bloomeries in Furness were suppressed in order that the woods might be no further destroyed. The reason for this is apparent when one remembers that it was at the time when the new Company of Mines Royal was being set on foot to exploit the mines in the Lake District for the benefit of the Crown; it was only after this monopoly had ceased that smelting was resumed at Hacket Forge in 1707.

I have no record of a stock of sheep here. The occupant would not have much time for sheep in the days of the forge, but the shippon has cow stalls so it is evident that they were kept in the past. There were a few fields which are now let off to another farm and the house is lived in as a cottage. The Wordsworths were familiar with the cottage, as was Lanty Slee. It was part of the Muncaster estate until 1942 when it was acquired by the mountaineer and author, the late Edmund W. Hodge of Elterwater Hall. The present owners bought the cottage in 1976.

Dale End

This property is approached on foot from the old Ullets Nest road, now in very bad repair, which joins the Elterwater to Colwith road on leaving

Elterwater village. This road continues and, after passing the farm, it soon reaches the main Little Langdale road which now gives vehicular access. The farm is situated by the roadside and contains parlour, back kitchen, living room, dairy, and four bedrooms. It is in a good sunny and elevated position with all the farm buildings nearby – it has approximately 124 acres of land and now belongs to the National Trust. The farm has fell grazing rights on Lingmoor and it used to have a stock of Herdwick sheep numbering 106 but this is now run by John Birkett of Wilson Place. The marks were forked near ear, under fold bitted both, and a stroke from the hock bone down far buttock and a pop on the tail head.

Thomas Harrison died at Dale End in 1661. Other occupants were in 1871 Edward Tyson who had three children. Later came Jonathan Birkett who was there when it was sold in 1936 as part of the Elterwater estate. After this it was farmed by William Redvers Hodgson who was baptised from High Hacket in 1900. It is now farmed by his younger son, William.

The Brow

This is a substantial stone and slated cottage which used to be a farm, occupying one of the finest sites in the district with commanding views over Little Langdale Tarn and up to Wetherlam. The small house contains living room, back kitchen, and two bedrooms. It has a few outbuildings, barn, loose box and peat house.

When it was sold in 1936 it was let to Jonathan Birkett along with Dale Head Farm which is quite near; now it is run as a holiday cottage.

I have a will of Robert Harrison of The Brow dated 1728. In it he mentions Thomas Harrison of Middlefell, William Harrison of Pye Howe, John Harrison of Wall End, William Sawrey late of Walthwaite, Edward Benson late of High Close, John Tyson of The Brow, John Grigg of Iving Howe, William Tyson of Johnson Place, and William Harrison of Field Foot, Loughrigg. The last two farms are now unknown.

Wilson Place

This farm is situated in a beautiful position in Little Langdale just north of the main road. It is a well-built house with two small cottages on the north end with the usual type of architecture. It contains five bedrooms, kitchen, and modern bathroom. Inside the sitting room, which was once the kitchen, is a lovely court cupboard with carved front and the initials DD 1645 IW which I understand are for Isaac Wilson and for his two daughters Deborah and Dora. His relatives developed Wilson's Kendal Mint Cake.

It has quite a good range of farm buildings together with appoximately 57 acres of land. It has grazing rights on Wrynose Common and is allowed 350 sheep and followers under the Registration Act and has a stock of

Herdwick sheep. The marks in the *Shepherd's Guide* of 1879 were cropped near ear and under key bitted far with a stroke from shoulder blade to hock bone on near side.

Wilson Place is one of the few farms not owned by the National Trust. It used to belong to the Elterwater estate and, when the estate was sold after the death of Mr Robinson in 1936, it was bought by Norman Birkett (later Lord Birkett), the judge. James Hodgson moved into the farm as tenant in 1948 and farmed it with his wife Molly, later buying the farm from Lord Birkett. When they retired it was taken over by James's nephew, John Birkett, who now farms it with his wife Maureen.

About 1790 William Stables lived at Wilson Place. He had five boys, Harry, who was born in 1809, my great grandfather, John, Edward, and William. The youngest was Jonathan, who was born in 1816, had a shop in Keswick, and died in 1877. In about 1870 John Harrison farmed here until he left in 1888 when James Braithwaite came. Later William Birkett farmed it, he had one son, Edward.

Recently it was necessary to renew the roof of the small, one-storey cottage at the entrance to Wilson Place. When the builders took off the old slates they found that they had been fixed with chicken bones. This is quite unique and the first time I have found bones to be used, the old-fashioned way was to fix them with wooden pegs.

Birk Howe, or Birch Howe

This farm is an old-fashioned and picturesque example of the local homesteads of the 17th century and still retains many of its original features. These include a kitchen fire-place of unique type, until it was modernised about 1947, beams so low that visitors avoid bumps only by using extreme care, low mullioned windows, oak-pegged doors, crude stone stairs, and bedrooms of extremely simple construction. It is in a beautiful position and well off the beaten track. The name comes from 'birk' (birch) and 'howe' (hill) – it seems that there have been birches standing here in the past. The farm buildings are all near the house, some actually attached. The land extends for about 95 acres and is situated in the middle of the valley, some of the fields bordering Little Langdale Tarn. A public footpath crosses the property on the way to Slater's Bridge – the old packhorse bridge.

The farm had belonged to the Youdell family for a very long while until Joseph Youdell sold it to Colonel Robinson, but later John Youdell bought it back again. When The Bield was sold he bought that also and his son John worked both these farms and his brother Jackson farmed Fell Foot.

In recent years it was farmed by Peter Hodgson, who had been a quarryman. He married Elizabeth Youdell in 1904 and they had a large family of four boys and three girls: James, George, a quarryman who married Bertha Hind, Peter, who married Mary Hind, and Jackson, and Dorothy, Mary, and Edith. He was 91 when he died in 1955. His sons James and Peter farmed here. When Peter married in 1939 he went to Fell Foot; James, who married Molly Nevison of Little Langdale in 1941, stayed until 1948. The National Trust bought the farm in 1947 since when it has been farmed by Frank Plaskett and then George Birkett.

It has a stock of sheep and the mark from the *Shepherd's Guide* of 1879 is cropped near ear, upper key bitted far, and a stroke down near shoulder and a pop on the tail head.

Low and High Bield

The Bield, known previously as Low Bield, is in a lovely setting but is no longer run as a farm. The house is quite large and contains some lovely oak beams and doors, black with age. It has a beautiful court cupboard, original, built into the wall with the initials and date, IIG 1714. There is another one carved with the inscription HB 1635. The house has no doubt been altered and some of the rooms made from a loft or barn. One of these was made into a studio for the artist Delmar Banner who lived here with his wife Josephina de Vasconcellos, the sculptress. In one large bedroom, Delmar painted a frieze all round the room – a beautiful panorama of the Lake District, beginning on one wall with the Langdale Pikes, Wetherlam, Crinkle Crags and Bowfell, then on to the Scafell range, Pillar, Grassmoor and round to Skiddaw. It has been a real labour of love. Josephine has also done several beautiful carvings. A lovely old porch with a sconson is still in evidence and some yew trees. The farm has a barn attached to the end.

Thomas Greenhow married Agnes Harrison, daughter of John Harrison of Wilson Place, in 1880 and took The Bield for about seven years when he left and went to Long House in 1887, and later Harry Place where he lived for the rest of his life. The Bield has not been farmed for a number of years, the land being let off to another farm. The house was bought by Fleming Mawson some years ago and was run as a family retreat before the Banners bought it. It is certainly a lovely dwelling place, well off the beaten track. It is now owned by Richard Bowness.

There was a stock of sheep at The Bield with the mark recorded in the *Shepherd's Guide* of 1879 punch holed near ear, and half crupper near side of the tail.

High Bield is a little higher up behind Low Bield; it is a small house with three bedrooms, now used as a cottage.

These two dwellings stand immediately under the Bield Borran – a safe place for foxes when being hunted. 'Bield' means shelter, for man or foxes. In fox hunting it signifies a fastness in the rocks which is inaccessible to the hounds.

At the opposite side of the entrance to The Bield is the field called Chapelmire. It is the site of the old chapel in Little Langdale possibly dating back to the 1600s. Some stones have been found which were perhaps part of the chapel. It seems quite a probable site as the road to Ravenglass via Wrynose and Hardknott is now believed to have run north of Little Langdale Tarn and it is fair to presume that the chapel was founded near the course of this road.

Busk

This is another old-world farm in a lovely position by the roadside overlooking Little Langdale Tarn and Wetherlam. It is about a quarter of a mile from the cattle grid, which used to be a gate at the bottom of Horse Brow. It is built in the traditional style with some lovely oak beams inside and contains four bedrooms. There is quite a good range of farm buildings but it has ceased to be a separate holding. It was last farmed by Harry Cowman until he left in 1963 to go to Stool End Farm. After this the land was put to other farms. It had a stock of sheep and the mark was cropped far, and a stroke from shoulder blade to back bone on far side.

In 1719 Robert Harrison of Pye Howe married Dorothy Vickers of Busk. Some other occupants were 1853 George Tyson, 1871 John Hodgson, followed by his son Ernest and after him Harry Cowman.

Fell Foot

This farm stands at the bottom of Wrynose Pass and is appopriately named Fell Foot. It is built in the usual style of Lakeland dwellings. There are no inscriptions to give us a clue about its date. Records show, however, that Sir Daniel Fleming's youngest son, Fletcher, bought it in 1707 and placed his coat of arms over the main door, high up over the gable.

Mr Palmer in his book *Historic Farmhouses in and around Westmorland* described the interior of the farmhouse in 1952:

> In the newer part of the farm, which may be 200 to 300 years old, some of the old features remain, despite modernisation at different times. A beautiful piece of work is the panelling in the principal bedroom, forming one of its walls. Over the fireplace is a plaster-work design of grapes and leaves, and traces of a frieze can also be seen round the walls. Leading from the centre room is a door to a tiny room which projects outwards over the front door. It is built of laths and plaster and had formerly a window to the front, and there are two seats on the inside. The window was fastened up when the slating was added to make it more weatherproof. Mrs.Youdell [the farmer's wife] could not explain what this tiny room was, and it looks too small for a spinning gallery. She had, however, a very good tale of what is said to have occurred here in the days when smuggling and the distilling of illicit whisky were common. The farm kept a keg of spirits here, and one day, when it was in danger of being found by the excise men, the lady of the house showed her presence of mind by sitting on the keg and covering it entirely with her crinoline, thus distracting their attention, while the men folk bored a hole through the floor and barrel and drained away the contents.

An early resident was Fletcher Fleming after he bought it in 1707 and it remained in the family of his descendents, the Flemings of Rayrigg, until it was sold to the National Trust in 1957. The Tysons were there as tenants for many years before it was farmed by Jackson Youdell and after that by Joseph Park. He was followed by Peter Hodgson, Jackson Youdell's nephew, who left in 1972 when Mr and Mrs George Harryman came.

The land is quite extensive having been extended by having the Bridge End and Blea Tarn land put to it. It has a stock of 337 sheep and they run up Greenburn and Wrynose. The sheep mark from the 1879 *Shepherd's Guide* was cropped both ears and a stroke over fillets and down both lisks – twinters a pop on near shoulder.

Bridge End, and High and Low Hallgarth

These were three old farm holdings in days gone by. They are over the beck from Fell Foot as you retrace your footsteps south. Over a stone bridge you reach Bridge End or Brig End. This dates back to at least 1578 and was a separate farm until some 50 years ago. There is some wood panelling in the living room; inscribed on one of the wooden beams is the date 1709. The last person to farm here was William Hodgson. After he left the land was given to Fell Foot and the house run as a holiday cottage owned by the National Trust.

Heading round the south side of the tarn we come to the two small farms of High and Low Hallgarth. High Hallgarth is a quaint little farm with outbuildings, perched on the edge of the slate quarry waste and now converted by the National Trust into holiday accommodation.

Low Hallgarth is run as a small farm but is too small to be viable unless the tenant does other outside work, which in the past was the case. In 1952 Alfred Webb, who was born in Langdale and had previously farmed Blea Tarn, lived at Low Hallgarth by himself and supplemented his living by buying and felling coppice wood and selling firewood. In 1952 a neighbour, Mr Plaskett, saw that there was no smoke coming from the chimney and went in and found him dead, sitting in his chair before a fireless grate. The house has since been occupied by his step-son, Herbert Thompson. Inside it has three bedrooms and a very large court cupboard in very good condition with the initials and date IHM 1682 (John and Margaret Holme) with nice carving.

CHAPTER 14

LOUGHRIGG FARMS

Loughrigg was an area of its own with its own lordship of the manor. As early as 1283 it was not mentioned by name but seemed to be included in Langdale; in a rental of 1375 Loughrigg was included in Langdale. For the purpose of the tithe apportionment on the map of 1838 it was entered as Loughrigg and Rydal township, but much of its history is linked with Langdale and this is a good reason for including it here.

Loughrigg Fold and Crag Head

One of the most important holdings was a group of houses or little hamlet called the Fold, now Loughrigg Fold, tucked away in a very sheltered situation under a small hill. On the top of this hill is the small pond or sheet of water called Tenters' Pond. The Benson family, who inhabited the houses at the Fold from the 16th century, were clothiers and wool merchants. This hamlet comprised a group of houses with the blacksmith's shop, the remains of which are still to be seen today near a huge chestnut tree. Near the road are the remains of a group of houses which were owned by George Law in 1797, the last one of which was depicted by William Green in a drawing published in 1822. Soon after this it was stripped of its oak fittings and demolished. Now you can still see the remains of the outer walls and some of the room patterns. No doubt the new barn was built with the stone.

There was a row of houses about the present Loughrigg Fold which adjoined the blacksmith's shop. Inside the present house is some beautiful old oak wood-work with a court cupboard built into the wall inscribed IB 1691 (John Benson). A stock of sheep was registered at Loughrigg Fold and the mark was ritted far ear and a pop on top of the shoulders and one on the fillets. Outside are some huge yew trees and a marvellous view of Loughrigg Tarn. Crag Head too has a lovely view and contains some nice old oak with a spice cupboard dated TSM 1708, probably for Thomas Sewart and his wife; at the east end of the house there are the remains of another cottage, now collapsed and covered with ivy. Loughrigg has been a beautiful little hamlet of which only these two houses remain.

This small hamlet was a Quaker stronghold, the Bensons being well-known members of the Society of Friends. Francis Benson with his wife Dorothy and John Dixon and William Harrison of Langdale were summoned to appear at the Quarter Sessions to answer the charge of having a meeting there; this was in 1663. The place of his burial is not known, though his death is recorded in April 1672 as being from the 'Fould in Loughrigg' in the Quaker register. There is a tradition of a burying ground somewhere about his vanished homestead and it is quite possible that some members of the family might be buried there. Meetings for worship would be held in one of the houses here and at least one wedding took place in December 1661.

The How

A gated road, called the Drive, leads round the back of Loughrigg Tarn. Here are two old small holdings, now lived in as cottages, and known as The How. These houses have a barn attached so have definitely been small farms in the past. The two places have magnificent views looking across Loughrigg Tarn with the Langdale Pikes in the background, and both get the sun all day.

Tarn Foot

If you continue along the Drive you come to Tarn Foot Farm, a very useful farm with a good selection of farm buildings. In 1885 it was run by James Mawson and later by John Allonby who had left Baysbrown in 1922, and then by his family. It is still in the family as Ann was John's great granddaughter and she married the present occupier, Malcolm Murphy.

High Close

In an old deed of 1621 Edward Benson was at High Close and he was selling Millbeck in Langdale to Robert Nicholson. The stock of sheep entered in the *Shepherd's Guide* of about 1800 shows that John Greenup lived in the farm and his sheep were marked under key bitted near ear, and upper halved far, and a red stroke over the fillets and down both sides. They were pastured on Bowfell in the summer.

In 1857 Edward Wheatley-Balme came over from Yorkshire and bought the farm off James Greenwood senior. He was later to buy a great part of Loughrigg, if not all of it. Having bought High Close he extended it into a large house and after this it ceased to be a farm, the land being put to other farms in the area. The *Shepherd's Guide* for 1879 shows that the High Close sheep mark was entered by Henry Mounsey of The Oaks Farm, so obviously the stock of sheep was put to The Oaks.

High Close is now run as a busy and successful Youth Hostel.

Scroggs

Coming down from High Close on the Skelwith road is a small holding called Scroggs, the house standing by the roadside in an attractive sunny position. It is an old-world farmhouse comprising living room, bathroom, and three bedrooms. Attached to the house is a barn with a shippon underneath and it would own some of the fields nearby for a few cows. I have no record of the farm having a stock of sheep. This small holding was bought by Mr Balme. It was used for one of his gardeners, Bernard Philipson, and is now occupied by his niece and family.

The Oaks

The Oaks was quite a large farm and no doubt most of the High Close land was put to this as it was adjoining, together with the sheep. Henry Mounsey, who was the tenant in 1879, was followed by his son Jackson who was there for about 50 years until he left in 1922, when his son Henry went to farm Low Millbeck. He was followed by the Wilsons and the Tysons, and by the Lucas, Emmit, and Nathan families. Mr Nathan died quite young of polio. Then Joe Park left Side House to farm there in 1948 and stayed until 1968 when it ceased to be a farm. The farm came to the National Trust, with the rest of the High Close estate, in lieu of death duties when Frank Edward Thorpe Jones-Balme died in 1951. The land was apportioned to the Langdale farmers. It was a sad day as it had been a really good farm.

The Ellers

Further down the road towards Skelwith is a small farm, comparatively new, having a datestone ECR 1849, called The Ellers. The farmhouse is on the roadside with the farm buildings opposite. These have now been made into holiday cottages. In the 1879 *Shepherd's Guide* it was run by Thomas Roberts and the sheep marks were slotted far ear, cropped near ear, and a large red 'T' on the far side. After this it was owned and farmed by John Roberts – he was there in 1885. It was later farmed by John Youdell and the last to run it as a farm was William Park. It then belonged to a Mr Horner whose daughter William had married. In 1961 the farm was sold, Parks left, and it ceased to be a farm, the land being put to other farms.

Mill Brow

Another old mill in this area is Mill Brow; evidence shows that a mill stood here two or three hundred years ago, driven by water flowing from Loughrigg Tarn. In wet weather it would have a good head of water. There were tenters here where the beck runs eventually into the Brathay. It was a customary holding of the Benson family.

The old farmstead has some very interesting features. There are some beautiful oak beams and a court cupboard inscribed HD1710. Above the little fire window and built into the wall is a Bible shelf or 'catmallison'; it is enclosed in oak – top, bottom, sides and back – but open to the room and measures on the inside 2 feet 9 inches x 4½ inches. In the same room is a simple spice cupboard to the left of the fireplace. On one of the flagstones in the floor of the living room is an inscription which is rather difficult to read:

Civil and religious liberty all over the world
The liberty of the press
If we have it not
T'is like the air we breathe – we die
Limited monarchy for ever

In later years it was owned by Thomas Preston – he lived there in 1885 and after him it was run by Miss Preston. Her farm-hand was Harry Wilson and her maid was Lizzie Greenhow, the daughter of Thomas Greenhow of Harry Place. After this it was owned by Mrs Manby and then by Miss Ainsworth and Miss Wood, who were there for 24 years, keeping a small herd of Jacob's sheep. It has recently been bought by George Long and his wife, an old Langdalian family.

In the 1879 *Shepherd's Guide* it was run by James Bousfield and the sheep mark was registered for Mill Brow and Brunt Howe as cropped far ear and a stroke down the far buttock and a pop on the near lisk.

Neaum Crag

'Not very far from Mill Brow is Neaum Crag. The residence of Albert Fleming is unique in construction. It is built into the face of the rock, and the only entrance to it is at the top. The views from its windows are said to surpass those obtained from any house in Westmorland' (Bulmer, *Westmorland Directory*, 1885). Albert Fleming was responsible for reviving the crafts of spinning and weaving in Langdale.

Mere Syke

Another interesting feature of the area is that a small mill may have stood near Mere Syke not far from a field called Long Brow. A building is known to have stood here, but has long since disappeared. The present house, Mere Syke, was at first called Woodlands and in 1885 was occupied by the artist Edward Tucker.

Skelwith Bridge Hotel

The present Skelwith Bridge Hotel was originally called the Hare and Hounds and has been greatly enlarged since. It was a small hotel with a farm attached together with a few fields.

On 27 May 1940 a meeting was called at the Co-op room in Chapel Stile to form the Local Defence Volunteers group which later became the Home Guard. A hundred and eight of us, drawn from the Langdales, Loughrigg, and Skelwith as far as Clappersgate, formed the No. 2 Platoon of C Company, 9 West Battalion. Thanks to John Kitchen, the owner of the Skelwith Bridge Hotel, we were able to have a guard room in an outbuilding which is now a cottage. We had bunk beds fitted, a few old chairs, and an open fire. Our main duty was to guard Skelwith Bridge and stop all cars and examine their documents. Although we were issued with uniforms on 18 October, we had no arms, only heavy rubber truncheons, until rifles were issued on 8 December. Not many of us had seen, let alone fired, a rifle when we were issued with a very heavy Lea Enfield 300 and bayonets! We practised at the old butts on Silverhowe and also in the Co-op room, shooting into a large barrel of sand.

We were on night duty one night in six; it was hard work being up all night and then having to return home to a day's work. When called out we had to use our push-bikes or walk in the dark. We had lessons in map-reading, which was a necessity as all the road signs had been taken down, in signalling, in first aid, and in the various poisonous gases.

One major excitement was the escape of German prisoners from Grizedale Hall. Possible identification was made more difficult due to large numbers of evacuees in the area. This period of my life was very interesting, we all learnt a lot and I enjoyed it all, making many good friends.

Loughrigg Fold

POSTSCRIPT

THE FUTURE – HOPES AND FEARS

About 25 years ago I forecast that the growing influx of tourists would eventually get the upper hand on the farmer. Damage is done by walkers knocking down walls and trespassing across one's fields damaging the grass and the crops. If you try to explain to them they often just say that they are just sitting on the grass – they are not damaging it; but I am sure that those same people would not want their bread and butter walking on! Invariably the picnickers leave a quantity of litter, or push paper, packets, and tins into the drystone walls. Surely it is not too much trouble for them to take their rubbish back with them. Twenty-odd years ago a 'litter warden' was never heard of – now there is a paid warden in most of the areas with a motor van provided – and all this costs money which could be spent in a much more worthwhile way.

Erosion is a very serious problem, but this is being taken care of by the upland management and voluntary wardens. Traffic is another problem. You hardly dare cross the road with sheep or cattle – you carry your life in your hands, and it is a worse hazard for dogs. During the latter part of my farming days I had two dogs killed by motor cars. This is very serious as good dogs are bad to find, and if you have one to buy, it can run into hundreds of pounds and take months to train. In these hill farming areas a good dog is a necessity – it would be impossible to gather your fell sheep without two or three dogs. I have seen them running on three legs and some with sore paws at gathering time.

We do welcome everybody to come and see our lovely countryside, but everyone has so much spare time and holidays. Years ago we were lucky if we got one week's holiday, now most people have two, three, or even four weeks, and each weekend. I really do not know the answer, but one will have to be found to prevent all the congestion with all the increase of the faster cars. In my young days 30mph was a fast speed – now cars are designed to travel at speeds up to 120mph. Traffic problems are particularly bad in Little Langdale and especially at farms like Fell Foot. If there was a fire or a serious illness, it could mean the difference between life or death. I am sure more passing places could be created without any scar on the landscape.

Changes in farming

Another big change has been the amalgamation of the farms. During my lifetime I could name about a dozen small farms that have disappeared. These small farms were a jumping-off place for young couples – there was not a lot of money to be made, but there were always little side-lines which could help out. The local life was much more sociable when I was young.

The methods of farming too are much different – now it is all done by mechanical means. You never see the old-fashioned hay rake. It is sad to see the hedges round the fields gradually disappearing and being replaced by wire fences. These hedges were planted on the top of a mound of earth and were a grand place for animals to shelter from the storm and also provided good cover for birds and small mammals. It is the policy now to make bigger fields for the much larger farm machinery. It is sad to see this change of farming practice but I suppose we have to move with the times. Admittedly many of the trees were suffering from old age and needed replanting. A big change to the farming methods was the arrival of the tractor in the valley about 40 years ago. One welcome implement was the muck-spreader. Before this the manure or 'muck' that was created by the horses and cattle was put into the midden and carted out periodically when the weather was suitable. It was all set out in heaps a certain distance apart, and then had to be scaled or spread by hand with the fork or 'gripe'. This was a back-breaking task – to start scaling a field of four or five acres really was a daunting task and I invariably went home at the end of the day with a very bad back. After it was scaled it had to be chain-harrowed to break it up better. Before the chain-harrow, we made our own brush-harrow with pieces of birch fastened together with string, rope or wire. The muck-spreader solved all this.

Nowadays there are a lot of artificial fertilisers used and these are helpful for flushing on the grass and crops but in many cases have been found to do a great deal of harm and be a cause of pollution. There is nothing like the good old-fashioned manure – now it's called 'organic farming' and is all coming to the fore again.

Homes and small businesses

There used to be many gentlemen's houses but these have now been turned into hostels or guest houses, now a necessity as more people come to see our lovely countryside. We now have a timeshare complex, I understand that it is one of the best in Europe – I don't think it is a blot on the landscape, and it provides locals with a lot of work in the valley – but 'enough is enough', and I hope it doesn't go too far. A bad thing is that many houses are bought up for second homes for people out of towns. This is unfortunate for the local young people, but they are lucky to have some new houses available only for local people, although the rents are pretty high as the costs of erecting the buildings are so large. I do think that the Lake District Planning Board does a good job – plans should be kept to a local style and I think it is a must that new buildings should be built and roofed in our local stone.

I would really like to see small businesses in the valley. In years gone by we had various crafts in the village. We had two or three grocer's shops, a chemist's, two or three shoemakers or cobblers, a clogger, a blacksmith, one or two tailors, joiner, wheelwright, and bakers, as well as spinners.

Unique beauty

We welcome all visitors, not just rock-climbers and fell-walkers but also sightseers and artists and old people to enjoy the scenery. Langdale Common is a beautiful place to sit and enjoy the lovely view. I'm sure that there is no more beautiful view in Britain or in the world – just to sit, if you are no longer able to walk on the fells. It is a pity that the National Trust has recently spread heaps of soil by the side of the road on the common. It would have been much nicer to make it possible for more cars to stop and for people to enjoy the view. We must not be selfish as more and more people come with such easy access to the Lakes. The National Trust do a splendid job but I think they have made a mistake in this case – they mustn't get carried away and think that all they do is right!

The mountain bikes have come and I'm sure that if they do no damage we should welcome them. We seem to get so much selfishness and some people who come to retire here think it is theirs for keeps, but this is not so. I have lived in Langdale all my life and have seen many changes – some for the better, some for the worse. But this is life and we must accept it. Many people will say that my remarks are not fully justifiable – so be it. However, these are my views and I'm glad to have been born and to have lived in such a beautiful part of the world. I have been fortunate to travel in British Columbia, in the Rockies, many times, and when I come back I can honestly say that there is no more beautiful place than Langdale.

WILD PLANTS FOR EATING AND REMEDIES FOR AILMENTS

Edible plants

Bilberry – a lovely fruit, we used to spend all day on Blea Rigg gathering them. The flavour of them is just something out of this world – they make a lovely tart!

Cranberry – in a few boggy places, not so common, but makes a lovely jelly.

Crab-apple – for making jelly.

Sloe – for wine-making or sloe gin.

Hazel nuts – we used to love going nutting in the autumn.

Beech nuts – not a big bite but very tasty.

Elderberry – for wine-making – 'Bawtry Jack'; the flowers make a wine too.

Hawthorn – berries for wine-making, jelly and syrup; the flowers are good for the bees.

Rowan – berries for wine.

Wild strawberries and raspberries – delicious.

Field mushrooms – frequent in certain places some years.

Pig nut or earth nut – (*conopodium majus*) – we used to dig up the roots and eat them when we were children, they were quite sweet to the taste.

Nettles – gathered in spring when young and treated like spinach, also as an ingredient in herb pudding.

Sorrel or sour dock – *rumex acetose* – can be used in many ways as a vegetable.

Mint – for making mint sauce, spearmint is best.

Hop shoots – (*humulus lupulus*) – look like asparagus, boil until tender, drain and serve with melted butter and lemon.

Juniper – three or four berries equivalent to one bay leaf.

Dandelion – blanched and used in salads.

Chives – not wild here but found in most gardens.

Watercress – found in many streams.

Good King Henry – grows by the roadside by Greenmire Field at the back of the Nick Stick Seat, used as a vegetable.

Rose hips – for making syrup.

Easterledge – the leaves are used in herb pudding.

Remedies

Honey – for sweetening and making mead, a very ancient drink.

Tar pills – we bought Stockholm tar, mixed it with flour, and made into balls as a treatment for piles – taken orally.

Goose grease – to rub on your chest.

Brimstone and treacle – we were given this as children if we had spots.

Carbolic soap – to guard against infections such as diphtheria or scarlet fever.

Camphor balls – to keep out moths.

Sheep's wool – for treating a corn – it contains plenty of lanoline.

LOCAL DIALECT

About twenty years ago a man from Norway visited and interviewed various Lakelanders who could speak the Westmorland dialect. He called on me and we spent some two hours together and he made a tape recording of our dialect words. After he had finished I asked him if he would say something in Norwegian, as our dialect is supposed to originate from the Norse settlers. He agreed and said that his wife would say to their children, 'Ga oot an lake' – meaning 'Go out and play', which it exactly what we would say!

Graham Hartley once said to me 'Oh Tom, what would Tommy Birkett mean when he said he was as weak as a turnip?' I told him that Tommy wouldn't have said that, but 'I'se as wake as a tewit' meaning 'I'm as weak as a tewit' or lapwing, as it flops about feigning injury attempting to lure away a predator from its nest or young. Some of these old sayings are very descriptive – the opposite to Tommy's remark would be 'as lish (active) as a cat'.

Proverbial sayings

Tak time in bank – Go steady as you take your horse and cart down a steep hill.
Yance (once) bitten, twice shy.
Time and patience a pig'll et a grindstean – pigs will eat stones and given time they would eat a grindstone!
When drinks in, wits oot.
A whistling woman and a crawn (crowing) hen sends t'devil to his den.
A new besom sweeps clean.
Bithr niver be born as Sunday shorn – should never have a haircut on Sunday.
Listeners niver hear good o' thersels.
Nowt as queer as folk, especially wick (lively) uns.
When t'winds in t'east, it's nadder good for man nor beast.
Good stuff's allus (always) in a lile (little) room – used in reply to a big boy who is boasting!
Taks a' macks to mack a world – takes all sorts to make a world.
Like fadder, like son.

Descriptive sayings

As sla (slow) as a haver (oatmeal) moo-at (mite or weevil).
As wisht (quiet) as a moose (mouse).
As tuff as bull lugs (ears).
As white as de-ath.
As black as auld Nick.
It caps cutlugs – meaning something unusual.
As black as Dick's hat bands.
A bigger leer than Tom Pepper!
A slaat (slate) off – something missing in the head.
Caat (can't) sit, stand, or hod bid gerse (hold a bit of grass) – used when someone is very drunk.

As de-ad as a do-or nail.
As blind (as in wind – short 'i') as a bat.
Soor as crab vargins – sour as crabapples.
As tuff as lidder (leather).
As slape (slippery) as an eel's tail.
As honest as day's leet (light).
As daft as two short planks.
As sharp as gerse leetning (greased lightening).
As thrang (busy) as Throp's wife.
As hard as brassel (brass).
Tight as wax – drunk.
As kittle (alert, delicately poised, sharp) as a moose trap.
Stinks ware (worse) eve'n a poak (purse) o' devils.
As straight as a yard of pump watter.
As bald as a peat (peewit or pate-head).
Moose missled (disappeared) off the face of the earth.
Moose was fisslen aboot (dashing about) in the bedroom.
Ower glisky – over bright to last (in the early morning).
Thoo gurt gammerstangs – stupid.
A' Ham Sam – mixed up.
Jannock – fair play.
Bait, bagging, or 10 o'clocks – food for mid-morning.
Bumbly kites – blackberries.
Fair brosson – out of breath (after climbing a hill).
Brossenfull – eaten too much.
Brant – steep.
Brat – an apron, a child.
Claggan – hard toffee made from treacle.
Cumd'y-gard'y – come day, go day, easy-going.
Fratch – fall out.
Geat – road, track, way.
Hag stock – chopping block.
Hag worms – slowworms.
Kysty – choosy, not keen on some foods.
Kopt – caught.
Kelterment – bits and pieces (in your pocket).
Late – look for.
Rowky – misty.
Riving – splitting.
Rovven – torn.
Scraffle – struggle.
Scrow – untidy – the house is in a scrow.

Farming terms

As rough as a bean stack.
A rapped sheep – when it sheds its wool, showing it has been in poor health during the winter.
A cowd tup – one without horns.
Cessen – a sheep when it cannot get up off its back.
Greand – sheep with upper jaw longer than the lower.
Reckling – runt, the weakest in the litter.

Scrat – sheep which is neither male nor female, deformed.
Assle tree – axle, cart-wheel bearing.
Bensle – to thrash.
Beasting – milk from a newly calved cow, makes a nice pudding like custard.
Cat-collop – tasty morsel out of a pig for the cat.
Chats or lile taties – little potatoes.
Coo-greup – the passage where the muck ligs (lies).
Fo moow – sunken mowings of hay, in the barn.
Outgang – track out from the farm to the fells.
Pooak – sack, purse, bag.
Scowp – ladle.
Stickle – spike.
Stoop – gate post.
Sype – drain off or sieve.

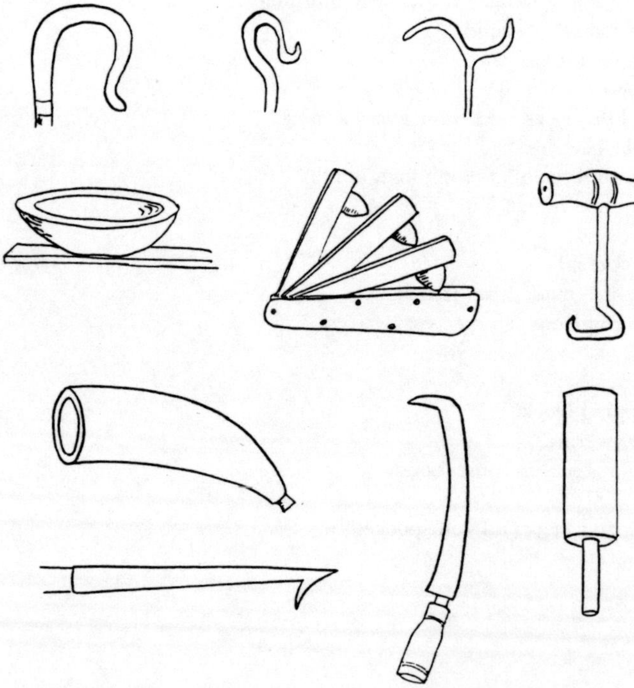

Shepherds crook	leg crook	dipping crook
Salving pan	Fleams	Sack or wool hook
Drenching horn	Turnip swagging tool	Blocking stick
Hay puller		

SOURCE MATERIAL

Printed books:

Armitt, M.L	1907 *Fullers and Freeholders of Grasmere.* Kendal.
	1912 *The Church of Grasmere.* Kendal.
	1916 *Rydal* (ed. W.F.Rawnsley). Kendal
Bradley, A.G	1901 *Highways and Byways in the Lake District.* London.
Bulmer, T.F.	1885 *History, topography and directory of Westmorland.* Preston.
Clapham, A.R. and others	1957 *Flora of the British Isles.* Cambridge.
Collingwood, W.G.	1906 *The Book of Coniston.* Kendal.
	1932 *The Lake Counties.* London.
Cook, O.	1954 *English Cottages and Farmhouses.* Norwich.
Cowper, H.S.	1897 *The Oldest Register Book of the Parish Hawkshead.* London.
Gate, D	1879 *Gate's New Shepherd's Guide.* Cockermouth.
Ewbank. J. (ed.)	1963 *Antiquary on Horseback.* Kendal.
Gibson, A.C.	1873 *Folk-speech of Cumberland.* London.
Kirkby, B.	1898 *Lakeland Words.* Kendal.
Machell, H.W.	1911 *Some Records of the Grasmere Sports.* Carlisle.
Nicolson, J. and Burn, R.	1777 *History ... of Westmoreland and Cumberland.* London.
Palmer, J.H.	1952 *Historic Farmhouses in and around Westmorland.* Kendal.
Pearsall, W.H. and Pennington, W.	1973 *The Lake District.* London.
Rollinson, W.	1974 *Life and Tradition in the Lake District.* London.
	1978 *A History of Cumberland and Westmorland.*London.
Winchester, A.J.	1987 *Landscape and Society in Medieval Cumbria.* Edinburgh.

MSS sources:

Census returns, land tax returns, jury verdicts, rental and survey 1573-74, Langdales and Loughrigg tithe apportionments and maps: Cumbria Record Office, Kendal, and Kendal Library.

Family collection: T.F. Buntin, now deposited in Cumbria Record Office, Kendal.